CURRICULUM AND EVALUATION

FOR SCHOOL MATHEMATICS

Prepared by the Working Groups of the
Commission on Standards for School Mathematics
of the National Council of Teachers of Mathematics

March 1989

NATIONAL COUNCIL OF
TEACHERS OF MATHEMATICS

The NCTM Commission on Standards for School Mathematics

Thomas A. Romberg, *Chair*
Iris M. Carl
F. Joe Crosswhite
John A. Dossey
James D. Gates
Shirley M. Frye
Shirley A. Hill
Christian R. Hirsch
Glenda Lappan
Dale Seymour
Lynn A. Steen
Paul R. Trafton
Norman Webb

Members of the Working Groups

K–4

Paul R. Trafton, *Chair*
Hilde Howden
Mary M. Lindquist
Edward C. Rathmell
Thomas E. Rowan
Charles S. Thompson

5–8

Glenda Lappan, *Chair*
Daniel T. Dolan
Joan F. Hall
Thomas E. Kieren
Judith E. Mumme
James E. Schultz

9–12

Christian R. Hirsch, *Chair*
Sue Ann McGraw
Gerald R. Rising
Harold L. Schoen
Cathy L. Seeley
Bert K. Waits

Evaluation

Norman Webb, *Chair*
Elizabeth Badger
Diane J. Briars
Thomas J. Cooney
Tej N. Pandey
Alba G. Thompson

Project Assistant

E. Anne Zarinnia

Library of Congress Cataloging-in-Publication Data

National Council of Teachers of Mathematics. Commission on Standards for School Mathematics.
Curriculum and evaluation standards for school mathematics / prepared by the working groups of the Commission on Standards for School Mathematics of the National Council of Teachers of Mathematics.
p. cm.
"June 1989."
Bibliography: p.
ISBN 0-87353-273-2
1. Mathematics—Study and teaching. 2. Curriculum evaluation.
I. Title.
QA11.N29 1989
510'.71—dc20

89-9421
CIP

Printed in the United States of America

Tenth printing 1996

TABLE OF CONTENTS

PREFACE

In 1986, the Board of Directors of the National Council of Teachers of Mathematics established the Commission on Standards for School Mathematics as one means to help improve the quality of school mathematics. This document, which is the product of the commission's efforts, contains a set of standards for mathematics curricula in North American schools (K–12) and for evaluating the quality of both the curriculum and student achievement. As school staffs, school districts, states, provinces, and other groups propose solutions to curricular problems and evaluation questions, these standards should be used as criteria against which their ideas can be judged.

The standards were drafted during the summer of 1987 and revised during the summer of 1988 by the members of four Working Groups, each representing a cross section of mathematics educators, including classroom teachers, supervisors, educational researchers, teacher educators, and university mathematicians (see page ii). They were appointed by John Dossey, president of NCTM. Their work was authorized and reviewed by the NCTM Commission on Standards for School Mathematics (see page ii), coordinated by Thomas A. Romberg and assisted by E. Anne Zarinnia. All who worked on the project thank the Utah Council of Teachers of Mathematics. Their hospitality both summers made our work easier and more enjoyable.

The revisions were based on copious and helpful reactions to the working draft of this document gathered during the 1987–88 school year. We thank all who contributed comments. This final document is considerably stronger and more coherent because of the careful reviews and thoughtful suggestions that were provided. In fact, we are confident this document represents the consensus of NCTM's members about the fundamental content that should be included in the school mathematics curriculum and about key issues concerning the organization and implementation of student and program evaluation.

The *Standards* is a document designed to establish a broad framework to guide reform in school mathematics in the next decade. In it a vision is given of what the mathematics curriculum should include in terms of content priority and emphasis. The challenge we issue to all interested in the quality of school mathematics is to work collaboratively to use these curriculum and evaluation standards as the basis for change so that the teaching and learning of mathematics in our schools is improved.

ACKNOWLEDGMENT

This document is significant because it expresses the consensus of professionals in the mathematical sciences for the direction of school mathematics in the next decade. It is also significant because it represents the total commitment of the National Council of Teachers of Mathematics to provide the leadership and resources for this vitally important work. The investment of time, money, and human efforts in the development of the *Curriculum and Evaluation Standards for School Mathematics* spans the terms of at least three presidents and boards of directors. Each one of us is proud to have been closely associated with the efforts.

We particularly acknowledge the outstanding leadership of Thomas A. Romberg, who chaired the Commission on Standards for School Mathematics, directed the writing project, and reported the deliberations to the profession. The Council is grateful to him, the members of the working groups, the project assistant, the Mathematical Sciences Education Board, and the thousands of other individuals and groups who offered their reactions to the initial draft.

Also, we give our special gratitude to the Headquarters staff, who supported the development and participated in the production throughout the entire process. All these activities were ultimately facilitated and guided by our executive director, James D. Gates, to whom we give our heartfelt appreciation.

F. Joe Crosswhite, President 1984–1986

John A. Dossey, President 1986–1988

Shirley M. Frye, President 1988–1990

ENDORSERS

The following mathematical science organizations join with the National Council of Teachers of Mathematics in promoting the vision of school mathematics described in the *Curriculum and Evaluation Standards for School Mathematics:*

American Mathematical Association of Two-Year Colleges
American Mathematical Society
American Statistical Association
Association for Women in Mathematics
Association of State Supervisors of Mathematics
Conference Board of the Mathematical Sciences
Council of Presidential Awardees in Mathematics
Council of Scientific Society Presidents
Institute of Management Sciences
Mathematical Association of America
Mathematical Sciences Education Board
National Council of Supervisors of Mathematics

Operations Research Society of America
School Science and Mathematics Association
Society for Industrial and Applied Mathematics

SUPPORTERS

The professional organizations listed below have added their support for the quality mathematics curricula and assessment criteria provided by the *Curriculum and Evaluation Standards for School Mathematics:*

American Association of Physics Teachers
American Association of School Administrators
American Chemical Society
American Federation of Teachers
Association for Supervision and Curriculum Development
Council for Basic Education
Council for Exceptional Children
Council of Chief State School Officers
Council of the Great City Schools
International Reading Association
International Technology Education Association
Junior Engineering Technical Society
National Association for the Education of Young Children
National Association of Biology Teachers
National Association of Elementary School Principals
National Association of Secondary School Principals
National Association of State Boards of Education
National Catholic Education Association
National Congress of Parents and Teachers
National Council for the Social Studies
National Council of Teachers of English
National Education Association
National School Boards Association
National Science Teachers Association
National Society of Professional Engineers

ALLIES

The organizations listed below have agreed to serve as allies in our effort to improve the teaching and learning of mathematics as described in the *Curriculum and Evaluation Standards for School Mathematics.*

American Association of Retired Persons
American Association of University Women
American Bankers Association
American Consulting Engineers Council
American Home Economics Association
American Indian Science and Engineering Society
American Institute of Certified Public Accountants
American Newspaper Publishers Association Foundation
Children's Television Workshop
Consumers Union
Indian Youth of America

Institute of Electrical and Electronics Engineers
Joint Council on Economic Education
Junior Achievement
National Coalition for Consumer Education
National Consumers League
National Council of LaRaza
National Council of Negro Women
National Federation of Business and Professional Women's Clubs
National Federation of Independent Business Foundation

INTRODUCTION

INTRODUCTION

Background

These standards are one facet of the mathematics education community's response to the call for reform in the teaching and learning of mathematics.[1] They reflect, and are an extension of, the community's responses to those demands for change.[2] Inherent in this document is a consensus that all students need to learn more, and often different, mathematics and that instruction in mathematics must be significantly revised.

As a function of NCTM's leadership in current efforts to reform school mathematics, the Commission on Standards for School Mathematics was established by the Board of Directors and charged with two tasks:

1. Create a coherent vision of what it means to be mathematically literate both in a world that relies on calculators and computers to carry out mathematical procedures and in a world where mathematics is rapidly growing and is extensively being applied in diverse fields.

2. Create a set of standards to guide the revision of the school mathematics curriculum and its associated evaluation toward this vision.

The Working Groups of the commission prepared the *Standards* in response to this charge.

This report is organized into six sections. This Introduction describes the need for standards, discusses the need for new goals, and presents an overview of the standards. The body of the report presents the standards themselves, organized into four distinct sections: K–4, 5–8, 9–12, and Evaluation. The concluding section outlines the steps necessary to accomplish the needed reform of school mathematics.

Key terms used in the development of this document include these three:

Curriculum. A curriculum is an operational plan for instruction that details what mathematics students need to know, how students are to achieve the identified curricular goals, what teachers are to do to help students develop their mathematical knowledge, and the context in which learning and teaching occur. In this document, the term describes what many would label as the "intended curriculum" or the "plan for a curriculum."

Evaluation. Standards have been articulated for evaluating both student performance and curricular programs, with an emphasis on the role of evaluative measures in gathering information on which teachers can base

---

[1]See *A Nation at Risk* (National Commission on Excellence in Education 1983) or *Educating Americans for the 21st Century* (National Science Board Commission on Precollege Education in Mathematics, Science and Technology 1983).

[2]*What is Fundamental and What is Not* (Conference Board of the Mathematical Sciences 1983a); *New Goals for Mathematical Sciences Education* (Conference Board of the Mathematical Sciences 1983b); and *School Mathematics: Options for the 1990's* (Romberg 1984).

♦ ♦ ♦ ♦ ♦ ♦ ♦ ♦

subsequent instruction. The standards also acknowledge the value of gathering information about student growth and achievement for research and administrative purposes.

Standard. A standard is a statement that can be used to judge the quality of a mathematics curriculum or methods of evaluation. Thus, standards are statements about what is valued.

The Need for Standards for School Mathematics

Historically there have been three reasons for groups to formally adopt a set of standards: (1) to ensure quality, (2) to indicate goals, and (3) to promote change. For NCTM, all three reasons are of equal importance.

First, standards often are used to ensure that the public is protected from shoddy products. For example, a druggist is not allowed to sell a drug unless it meets certain very rigid standards that include both the control of how it was produced and evidence of its effectiveness. Standards in this sense are minimal criteria for quality. They set necessary, but not sufficient, conditions for producing desired results. There is no guarantee that a drug will not be misused or will produce expected results.

Second, standards often are used as a means of expressing expectations about goals. Goals are broad statements of social intent. For example, we can agree that two goals for all tests are that they should be both valid and reliable. The standards for tests developed by the American Psychological Association in 1974 describe the kind of documentation that publishers should provide about the reliability and validity of each test.

Third, standards often are set to lead a group toward some new desired goals. For example, the medical profession has adopted and periodically updates standards for the licensing of specialists based on changes in technology, research, and so on. The intent is to improve or update practices when necessary. In this sense, standards should be seen as "criteria for excellence." They are based on an informed vision of what should be done given current knowledge and experience.

Standards are needed for school mathematics for all three purposes. Schools, teachers, students, and the public at large currently enjoy no protection from shoddy products. It seems reasonable that anyone developing products for use in mathematics classrooms should document how the materials are related to current conceptions of what content is important to teach and should present evidence about their effectiveness. For NCTM the development of standards as statements of criteria for excellence in order to produce change was the focus. Schools, and in particular school mathematics, must reflect the important consequences of the current reform movement if our students are to be adequately prepared to live in the twenty-first century. The standards should be viewed as facilitators of reform.

The Need for New Goals

Our vision of mathematical literacy is based on a reexamination of educational goals. Historically, societies have established schools to—

- transmit aspects of the culture to the young;
- direct students toward, and provide them with, an opportunity for self-fulfillment.

Thus, the goals all schools try to achieve are both a reflection of the needs of society and the needs of students.

Calls for reform in school mathematics suggest that new goals are needed. All industrialized countries have experienced a shift from an industrial to an information society, a shift that has transformed both the aspects of mathematics that need to be transmitted to students and the concepts and procedures they must master if they are to be self-fulfilled, productive citizens in the next century.

The Information Society. This social and economic shift can be attributed, at least in part, to the availability of low-cost calculators, computers, and other technology. The use of this technology has dramatically changed the nature of the physical, life, and social sciences; business; industry; and government. The relatively slow mechanical means of communication—the voice and the printed page—have been supplemented by electronic communication, enabling information to be shared almost instantly with persons—or machines—anywhere. Information is the new capital and the new material, and communication is the new means of production. The impact of this technological shift is no longer an intellectual abstraction. It has become an economic reality. Today, the pace of economic change is being accelerated by continued innovation in communications and computer technology.

New Societal Goals. Schools, as now organized, are a product of the industrial age. In most democratic countries, common schools were created to provide most youth the training needed to become workers in fields, factories, and shops. As a result of such schooling, students also were expected to become literate enough to be informed voters. Thus, minimum competencies in reading, writing, and arithmetic were expected of all students, and more advanced academic training was reserved for the select few. These more advantaged students attended the schools that were expected to educate the future cultural, academic, business, and government leaders.

The educational system of the industrial age does not meet the economic needs of today. New social goals for education include (1) mathematically literate workers, (2) lifelong learning, (3) opportunity for all, and (4) an informed electorate. Implicit in these goals is a school system organized to serve as an important resource for all citizens throughout their lives.

1. *Mathematically literate workers.* The economic status quo in which factory employees work the same jobs to produce the same goods in the same manner for decades is a throwback to our industrial-age past. Today, economic survival and growth are dependent on new factories established to produce complex products and services with very short market cycles. It is a literal reality that before the first products are sold, new replacements are being designed for an ever-changing market. Concurrently, the research division is at work developing new ideas to feed to the design groups to meet the continuous clamor for new products that are, in turn, channeled into the production arena. Traditional notions of basic mathematical competence have been outstripped by ever-higher expectations of the skills and knowledge of workers; new methods of production demand a technologically competent work force. The U.S. Congressional Office of Technology Assessment (1988) claims that employees must be prepared to understand the complexities and technologies of communication, to ask questions, to assimilate unfamiliar information, and to work cooperatively in teams. Businesses no longer seek workers with strong backs, clever hands, and "shopkeeper" arithmetic skills. In fact, it is claimed that the "most significant growth in new jobs

between now and the year 2000 will be in fields requiring the most education" (Lewis 1988, p. 468). Henry Pollak (1987), a noted industrial mathematician, recently summarized the mathematical expectations for new employees in industry:

- The ability to set up problems with the appropriate operations
- Knowledge of a variety of techniques to approach and work on problems
- Understanding of the underlying mathematical features of a problem
- The ability to work with others on problems
- The ability to see the applicability of mathematical ideas to common and complex problems
- Preparation for open problem situations, since most real problems are not well formulated
- Belief in the utility and value of mathematics

Notice the difference between the skills and training inherent in these expectations and those acquired by students working independently to solve explicit sets of drill and practice exercises. Although mathematics is not taught in schools solely so students can get jobs, we are convinced that in-school experiences reflect to some extent those of today's workplace. This is especially true given that the availability of such broadly educated workers will be a major factor in determining how businesses respond to today's changing economic conditions.

2. *Lifelong learning.* Employment counselors, cognizant of the rapid changes in technology and employment patterns, are claiming that, on average, workers will change jobs at least four to five times during the next twenty-five years and that each job will require retraining in communication skills. Thus, a flexible workforce capable of lifelong learning is required; this implies that school mathematics must emphasize a dynamic form of literacy. Problem solving—which includes the ways in which problems are represented, the meanings of the language of mathematics, and the ways in which one conjectures and reasons—must be central to schooling so that students can explore, create, accommodate to changed conditions, and actively create new knowledge over the course of their lives.

3. *Opportunity for all.* The social injustices of past schooling practices can no longer be tolerated. Current statistics indicate that those who study advanced mathematics are most often white males. Women and most minorities study less mathematics and are seriously underrepresented in careers using science and technology. Creating a just society in which women and various ethnic groups enjoy equal opportunities and equitable treatment is no longer an issue. Mathematics has become a critical filter for employment and full participation in our society. We cannot afford to have the majority of our population mathematically illiterate: Equity has become an economic necessity.

4. *Informed electorate.* In a democratic country in which political and social decisions involve increasingly complex technical issues, an educated, informed electorate is critical. Current issues—such as environmental protection, nuclear energy, defense spending, space exploration, and tax-

ation—involve many interrelated questions. Their thoughtful resolution requires technological knowledge and understanding. In particular, citizens must be able to read and interpret complex, and sometimes conflicting, information.

In summary, today's society expects schools to insure that all students have an opportunity to become mathematically literate, are capable of extending their learning, have an equal opportunity to learn, and become informed citizens capable of understanding issues in a technological society. As society changes, so must its schools.

New Goals for Students. Educational goals for students must reflect the importance of mathematical literacy. Toward this end, the K–12 standards articulate five general goals for all students: (1) that they learn to value mathematics, (2) that they become confident in their ability to do mathematics, (3) that they become mathematical problem solvers, (4) that they learn to communicate mathematically, and (5) that they learn to reason mathematically. These goals imply that students should be exposed to numerous and varied interrelated experiences that encourage them to value the mathematical enterprise, to develop mathematical habits of mind, and to understand and appreciate the role of mathematics in human affairs; that they should be encouraged to explore, to guess, and even to make and correct errors so that they gain confidence in their ability to solve complex problems; that they should read, write, and discuss mathematics; and that they should conjecture, test, and build arguments about a conjecture's validity.

The opportunity for all students to experience these components of mathematical training is at the heart of our vision of a quality mathematics program. The curriculum should be permeated with these goals and experiences so that they become commonplace in the lives of students. We are convinced that if students are exposed to the kinds of experiences outlined in the *Standards*, they will gain *mathematical power.* This term denotes an individual's abilities to explore, conjecture, and reason logically, as well as the ability to use a variety of mathematical methods effectively to solve nonroutine problems. This notion is based on the recognition of mathematics as more than a collection of concepts and skills to be mastered; it includes methods of investigating and reasoning, means of communication, and notions of context. In addition, for each individual, mathematical power involves the development of personal self-confidence.

Toward this end, we see classrooms as places where interesting problems are regularly explored using important mathematical ideas. Our premise is that *what* a student learns depends to a great degree on *how* he or she has learned it. For example, one could expect to see students recording measurements of real objects, collecting information and describing their properties using statistics, and exploring the properties of a function by examining its graph. This vision sees students studying much of the same mathematics currently taught but with quite a different emphasis; it also sees some mathematics being taught that in the past has received little emphasis in schools.

1. *Learning to value mathematics.* Students should have numerous and varied experiences related to the cultural, historical, and scientific evolution of mathematics so that they can appreciate the role of mathematics in the development of our contemporary society and explore relationships among mathematics and the disciplines it serves: the physical and life sciences, the social sciences, and the humanities.

Throughout the history of mathematics, practical problems and theoretical pursuits have stimulated one another to such an extent that it is impossible to disentangle them. Even today, as theoretical mathematics has burgeoned in its diversity and deepened in its complexity and abstraction, it has become more concrete and vital to our technologically oriented society. It is the intent of this goal—learning to value mathematics—to focus attention on the need for student awareness of the interaction between mathematics and the historical situations from which it has developed and the impact that interaction has on our culture and our lives.

2. *Becoming confident in one's own ability.* As a result of studying mathematics, students need to view themselves as capable of using their growing mathematical power to make sense of new problem situations in the world around them. To some extent, everybody is a mathematician and does mathematics consciously. To buy at the market, to measure a strip of wallpaper, or to decorate a ceramic pot with a regular pattern is doing mathematics. School mathematics must endow all students with a realization that doing mathematics is a common human activity. Having numerous and varied experiences allows students to trust their own mathematical thinking.

3. *Becoming a mathematical problem solver.* The development of each student's ability to solve problems is essential if he or she is to be a productive citizen. We strongly endorse the first recommendation of *An Agenda for Action* (National Council of Teachers of Mathematics 1980): "Problem solving must be the focus of school mathematics" (p. 2). To develop such abilities, students need to work on problems that may take hours, days, and even weeks to solve. Although some may be relatively simple exercises to be accomplished independently, others should involve small groups or an entire class working cooperatively. Some problems also should be open-ended with no right answer, and others need to be formulated.

4. *Learning to communicate mathematically.* The development of a student's power to use mathematics involves learning the signs, symbols, and terms of mathematics. This is best accomplished in problem situations in which students have an opportunity to read, write, and discuss ideas in which the use of the language of mathematics becomes natural. As students communicate their ideas, they learn to clarify, refine, and consolidate their thinking.

5. *Learning to reason mathematically.* Making conjectures, gathering evidence, and building an argument to support such notions are fundamental to doing mathematics. In fact, a demonstration of good reasoning should be rewarded even more than students' ability to find correct answers.

In summary, the intent of these goals is that students will become mathematically literate. This term denotes an individual's ability to explore, to conjecture, and to reason logically, as well as to use a variety of mathematical methods effectively to solve problems. By becoming literate, their mathematical power should develop.

An Overview of the Curriculum and Evaluation Standards

This document presents fifty-four standards divided among four categories: grades K–4, 5–8, 9–12, and evaluation. The four categories are arbitrary in that they are not intended to reflect school structure; in fact, we encourage readers to consider these as K–12 standards. In addition, we believe that similar standards need to be developed for both preschool programs and those beyond high school.

It was our task to prepare the curriculum and evaluation standards that reflect our vision of how the societal and student goals already articulated here could be met. These standards should be seen as an initial step in the lengthy process of bringing about reform in school mathematics.

Curriculum Standards. When a set of curricular standards is specified for school mathematics, it should be understood that the standards are value judgments based on a broad, coherent vision of schooling derived from several factors: societal goals, student goals, research on teaching and learning, and professional experience. Each standard starts with a statement of what mathematics the curriculum should include. This is followed by a description of the student activities associated with that mathematics and a discussion that includes instructional examples.

Mathematics. The first consideration in preparing each standard was its mathematical content. To decide on what is fundamental in so vast and dynamic a discipline as mathematics is no easy task. John Dewey's (1916) distinction between "knowledge" and the "record of knowledge" may clarify this point. For many, "to know" means to identify the basic concepts and procedures of the discipline. For many nonmathematicians, arithmetic operations, algebraic manipulations, and geometric terms and theorems constitute the elements of the discipline to be taught in grades K–12. This may reflect the mathematics they studied in school or college rather than a clear insight into the discipline itself.

Three features of mathematics are embedded in the Standards. First, "knowing" mathematics is "doing" mathematics. A person gathers, discovers, or creates knowledge in the course of some activity having a purpose. This active process is different from mastering concepts and procedures. We do not assert that informational knowledge has no value, only that its value lies in the extent to which it is useful in the course of some purposeful activity. It is clear that the fundamental concepts and procedures from some branches of mathematics should be known by all students; established concepts and procedures can be relied on as fixed variables in a setting in which other variables may be unknown. But instruction should persistently emphasize "doing" rather than "knowing that."

Second, some aspects of doing mathematics have changed in the last decade. The computer's ability to process large sets of information has made quantification and the logical analysis of information possible in such areas as business, economics, linguistics, biology, medicine, and sociology. Change has been particularly great in the social and life sciences. In fact, quantitative techniques have permeated almost all intellectual disciplines. However, the fundamental mathematical ideas needed in these areas are not necessarily those studied in the traditional algebra-geometry-precalculus-calculus sequence, a sequence designed with engineering and physical science applications in mind. Because mathematics is a foundation discipline for other disciplines and grows in direct proportion to its utility, we believe that the curriculum for all students must provide opportunities to develop an understanding of mathematical models, structures, and simulations applicable to many disciplines.

Third, changes in technology and the broadening of the areas in which mathematics is applied have resulted in growth and changes in the discipline of mathematics itself. Davis and Hersh (1981) claim that we are now in a golden age of mathematical production, with more than half of

◆ ◆ ◆ ◆ ◆ ◆ ◆ ◆

all mathematics having been invented since World War II. In fact, they argue that "there are two inexhaustible sources of new mathematical questions. One source is the development of science and technology, which make ever new demands on mathematics for assistance. The other source is mathematics itself . . . each new, completed result becomes the potential starting point for several new investigations" (p. 25). The new technology not only has made calculations and graphing easier, it has changed the very nature of the problems important to mathematics and the methods mathematicians use to investigate them. Because technology is changing mathematics and its uses, we believe that—

- appropriate calculators should be available to all students at all times;
- a computer should be available in every classroom for demonstration purposes;
- every student should have access to a computer for individual and group work;
- students should learn to use the computer as a tool for processing information and performing calculations to investigate and solve problems.

We recognize, however, that access to this technology is no guarantee that any student will become mathematically literate. Calculators and computers for users of mathematics, like word processors for writers, are tools that simplify, but do not accomplish, the work at hand. Thus, our vision of school mathematics is based on the fundamental mathematics students will need, not just on the technological training that will facilitate the use of that mathematics.

Similarly, the availability of calculators does not eliminate the need for students to learn algorithms. Some proficiency with paper-and-pencil computational algorithms is important, but such knowledge should grow out of the problem situations that have given rise to the need for such algorithms. Furthermore, when one needs to calculate to find an answer to a problem, one should be aware of the choices of methods (see fig. 1). When an approximate answer is adequate, one should estimate. If a precise answer is needed, an appropriate procedure must be chosen. Many problems should be solved by mental calculation (multiplying by 10, taking half). Some calculations, if not too complex, should be solved by following standard paper-and-pencil algorithms. For more complex calculations, the calculator should be used (column addition, long division). And finally, if many iterative calculations are required, a computer program should be written or used to find answers (finding a sum of squares). Note in figure 1 that estimation can, and should, be used in conjunction with procedures yielding exact answers to foreshadow any calculation and to judge the reasonableness of results.

Contrary to the fears of many, the availability of calculators and computers has expanded students' capability of performing calculations. There is no evidence to suggest that the availability of calculators makes students dependent on them for simple calculations. Students should be able to decide when they need to calculate and whether they require an exact or approximate answer. They should be able to select and use the most appropriate tool. Students should have a balanced approach to calculation, be able to choose appropriate procedures, find answers, and judge the validity of those answers.

Finally, in developing the standards, we considered the content appropri-

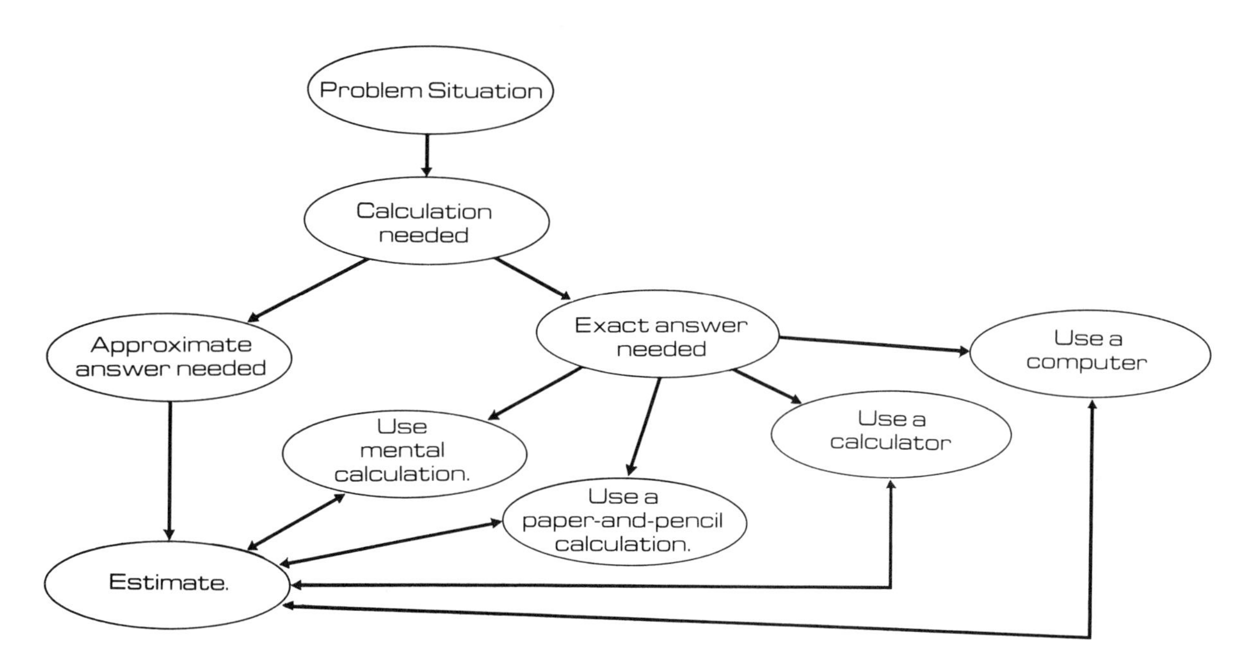

Fig. 1. Decisions about calculation procedures in numerical problems

ate for *all* students. This, however, does not suggest that we believe all students are alike. We recognize that students exhibit different talents, abilities, achievements, needs, and interests in relationship to mathematics. The mathematical content outlined in the Standards is what we believe all students will need if they are to be productive citizens in the twenty-first century. If all students do not have the opportunity to learn this mathematics, we face the danger of creating an intellectual elite and a polarized society. The image of a society in which a few have the mathematical knowledge needed for the control of economic and scientific development is not consistent either with the values of a just democratic system or with its economic needs.

We believe that all students should have an opportunity to learn the important ideas of mathematics expressed in these standards. On the one hand, prior to grade 9, we have refrained from specifying alternative instructional patterns that would be consistent with our vision. On the other hand, for grades 9–12, the standards have been prepared in light of a core program for all students, with explicit differentiation in terms of depth and breadth of treatment and the nature of applications for college-bound students. At the same time, the mathematics of the core program is sufficiently broad and deep so that students' options for further study would not be limited. Our expectation is that all students must have an opportunity to encounter typical problem situations related to important mathematical topics. However, their experiences may differ in the vocabulary or notations used, the complexity of arguments, and so forth.

Student Activities. The second aspect of each standard specifies the *expected student activities* associated with doing mathematics. Two general principles have guided our descriptions: First, activities should grow out of problem situations; and second, learning occurs through active as well as passive involvement with mathematics.

Traditional teaching emphases on practice in manipulating expressions and practicing algorithms as a precursor to solving problems ignore the fact that knowledge often emerges from the problems. This suggests that instead of the expectation that skill in computation should precede word problems, experience with problems helps develop the ability to compute. Thus, present strategies for teaching may need to be reversed; knowledge often should emerge from experience with problems. In this way, students may recognize the need to apply a particular concept or procedure and have a strong conceptual basis for reconstructing their knowledge at a later time.

Furthermore, students need to experience genuine problems regularly. A genuine problem is a situation in which, for the individual or group concerned, one or more appropriate solutions have yet to be developed. The situation should be complex enough to offer challenge but not so complex as to be insoluble. In sum, we believe that learning should be guided by the search to answer questions—first at an intuitive, empirical level; then by generalizing; and finally by justifying (proving).

In many classrooms, learning is conceived of as a process in which students passively absorb information, storing it in easily retrievable fragments as a result of repeated practice and reinforcement. Research findings from psychology indicate that learning does not occur by passive absorption alone (Resnick 1987). Instead, in many situations individuals approach a new task with prior knowledge, assimilate new information, and construct their own meanings. For example, before young children are taught addition and subtraction, they can already solve most addition and subtraction problems using such routines as "counting on" and "counting back" (Romberg and Carpenter 1986). As instruction proceeds, children often continue to use these routines in spite of being taught more formal problem-solving procedures. They will accept new ideas only when their old ideas do not work or are inefficient. Furthermore, ideas are not isolated in memory but are organized and associated with the natural language that one uses and the situations one has encountered in the past. This constructive, active view of the learning process must be reflected in the way much of mathematics is taught. Thus, instruction should vary and include opportunities for—

- appropriate project work;
- group and individual assignments;
- discussion between teacher and students and among students;
- practice on mathematical methods;
- exposition by the teacher.

Our ideas about problem situations and learning are reflected in the verbs we use to describe student actions (e.g., to investigate, to formulate, to find, to verify) throughout the *Standards*.

Focus and Discussion. Finally, our vision sees teachers encouraging students, probing for ideas, and carefully judging the maturity of a student's thoughts and expressions. Hence, each standard is elaborated on in a *Focus* section followed by a discussion with examples, which is meant to convey the spirit of this vision about both mathematical content and instruction.

Another premise of the standards is that problem situations must keep pace with the maturity—both mathematical and cultural—and experience of the students. For example, the primary grades should emphasize the empirical language of the mathematics of whole numbers, common fractions, and descriptive geometry. In the middle grades, empirical mathematics should be extended to other numbers, and the emphasis should shift to building the abstract language of mathematics needed for algebra and other aspects of mathematics. High school mathematics should emphasize functions, their representations and uses, modeling, and deductive proofs.

The standards specify that instruction should be developed from problem situations. As long as the situations are familiar, conceptions are created from objects, events, and relationships in which operations and strategies are well understood. In this way, students develop a framework of support that can be drawn upon in the future, when rules may well have been forgotten but the structure of the situation remains embedded in the memory as a foundation for reconstruction. Situations should be sufficiently simple to be manageable but sufficiently complex to provide for diversity in approach. They should be amenable to individual, small-group, or large-group instruction, involve a variety of mathematical domains, and be open and flexible as to the methods to be used.

The first three standards in each section are labeled Problem Solving, Communication, and Reasoning, although details vary between levels with respect to what is expected both of students and of instruction. This variation reflects the developmental level of the students, their mathematical background, and the specific mathematical content.

The fourth curriculum standard at each level is titled Mathematical Connections. This label emphasizes our belief that although it is often necessary to teach specific concepts and procedures, mathematics must be approached as a whole. Concepts, procedures, and intellectual processes are interrelated. In a significant sense, "the whole is greater than the sum of its parts." Thus, the curriculum should include deliberate attempts, through specific instructional activities, to connect ideas and procedures both among different mathematical topics and with other content areas. Following the Connections standard, nine or ten specific content standards are stated and discussed. Some have similar titles, which reflects that a content area needs emphasis across the curriculum; however, once again the concepts and processes emphasized vary by level. Others emphasize specific content that needs to be developed at that level.

The Evaluation Standards. The evaluation standards are presented separately, not because evaluation should be separated from the curriculum, but because planning for the gathering of evidence about student and program outcomes is different. The difference is most clearly illustrated in comparing the curriculum standards titled Connections and the evaluation standards titled Mathematical Power. Both deal with connections among concepts, procedures, and intellectual methods, but the curriculum standards are related to the instructional plan whereas the evaluation standards address the ways in which students integrate these connections intellectually so that they develop mathematical power.

We present fourteen evaluation standards that can be viewed in three categories. The first set of three evaluation standards discusses general assessment strategies related to the curriculum standards. The second seven focus on providing information to teachers for instructional purposes. They closely parallel the curriculum standards—problem solving, communication, reasoning, mathematical concepts, and mathematical procedures, in addition to a separate standard on "mathematical disposition." These seven standards are to be used by teachers to make judgments about students and their mathematical progress. The final set of four standards addresses the gathering of evidence with respect to the quality of the mathematics program. These standards are to be used by teachers, administrators, and policy makers to make judgments about the quality of the mathematics program and the effectiveness of instruction.

Challenge

Such are the background, the general focus, and the intent of our efforts. It is now left to each of you concerned with the teaching and learning of mathematics to read the standards, to share them with colleagues, and to reflect on their vision. Consider what needs to be done and what you can do, and collaborate with others to implement the standards for the benefit of our students, as well as for our social and economic future.

GRADES K—4

♦ ♦ ♦ ♦ ♦ ♦ ♦ ♦

CURRICULUM STANDARDS FOR GRADES K–4

OVERVIEW

This section presents thirteen curriculum standards for grades K–4:

1. ***Mathematics as Problem Solving***
2. ***Mathematics as Communication***
3. ***Mathematics as Reasoning***
4. ***Mathematical Connections***
5. ***Estimation***
6. ***Number Sense and Numeration***
7. ***Concepts of Whole Number Operations***
8. ***Whole Number Computation***
9. ***Geometry and Spatial Sense***
10. ***Measurement***
11. ***Statistics and Probability***
12. ***Fractions and Decimals***
13. ***Patterns and Relationships***

The Need for Change

The need for curricular reform in K–4 mathematics is clear. Such reform must address both the content and emphasis of the curriculum as well as approaches to instruction. A long-standing preoccupation with computation and other traditional skills has dominated both *what* mathematics is taught and *the way* mathematics is taught at this level. As a result, the present K–4 curriculum is narrow in scope; fails to foster mathematical insight, reasoning, and problem solving; and emphasizes rote activities. Even more significant is that children begin to lose their belief that learning mathematics is a sense-making experience. They become passive receivers of rules and procedures rather than active participants in creating knowledge.

The Direction of Change

The Introduction describes a vision for school mathematics built around five overall curricular goals for students to achieve: learning to value mathematics, becoming confident in one's own ability, becoming a mathematical problem solver, learning to communicate mathematically, and learning to reason mathematically. This vision addresses what mathematics is, what it means to know and do mathematics, what teachers should do when they teach mathematics, and what children should do when they learn mathematics. The K–4 standards reflect the implications of this vision for the curriculum in the early grades and present a coherent viewpoint about mathematics, about children, and about the learning of mathematics by children.

Children and Mathematics: Implications for the K–4 Curriculum

An appropriate curriculum for young children that reflects the *Standards'* overall goals must do the following:

1. *Address the relationship between young children and mathematics.* Children enter kindergarten with considerable mathematical experience, a partial understanding of many concepts, and some important skills, including counting. Nonetheless, it takes careful planning to create a curriculum that capitalizes on children's intuitive insights and language in selecting and teaching mathematical ideas and skills. It is clear that children's intellectual, social, and emotional development should guide the kind of mathematical experiences they should have in light of the overall goals for learning mathematics. The notion of a *developmentally appropriate* curriculum is an important one.

A developmentally appropriate curriculum encourages the exploration of a wide variety of mathematical ideas in such a way that children retain their enjoyment of, and curiosity about, mathematics. It incorporates real-world contexts, children's experiences, and children's language in developing ideas. It recognizes that children need considerable time to construct sound understandings and develop the ability to reason and communicate mathematically. It looks beyond what children appear to know to determine how they think about ideas. It provides repeated contact with important ideas in varying contexts throughout the year and from year to year.

Programs that provide limited developmental work, that emphasize symbol manipulation and computational rules, and that rely heavily on paper-and-pencil worksheets do not fit the natural learning patterns of children and do not contribute to important aspects of children's mathematical development.

2. *Recognize the importance of the qualitative dimensions of children's learning.* The mathematical ideas that children acquire in grades K–4 form the basis for all further study of mathematics. Although quantitative considerations have frequently dominated discussions in recent years, qualitative considerations have greater significance. Thus, how well children come to understand mathematical ideas is far more important than how many skills they acquire. The success with which programs at later grade levels achieve their goals depends largely on the quality of the foundation that is established during the first five years of school.

3. *Build beliefs about what mathematics is, about what it means to know and do mathematics, and about children's view of themselves as*

mathematics learners. The beliefs that young children form influence not only their thinking and performance during this time but also their attitude and decisions about studying mathematics in later years. Beliefs also become more resistant to change as children grow older. Thus, affective dimensions of learning play a significant role in, and must influence, curriculum and instruction.

ASSUMPTIONS

Several basic assumptions governed the selection and shaping of the K–4 standards.

1. *The K–4 curriculum should be conceptually oriented.* The view that the K–4 curriculum should emphasize the development of mathematical understandings and relationships is reflected in the discussions about the content and emphasis of the curriculum. A conceptual approach enables children to acquire clear and stable concepts by constructing meanings in the context of physical situations and allows mathematical abstractions to emerge from empirical experience. A strong conceptual framework also provides anchoring for skill acquisition. Skills can be acquired in ways that make sense to children and in ways that result in more effective learning. A strong emphasis on mathematical concepts and understandings also supports the development of problem solving.

Emphasizing mathematical concepts and relationships means devoting substantial time to the development of understandings. It also means relating this knowledge to the learning of skills by establishing relationships between the conceptual and procedural aspects of tasks. The time required to build an adequate conceptual base should cause educators to rethink when children are expected to demonstrate a mastery of complex skills. A conceptually oriented curriculum is consistent with the overall curricular goals in this report and can result in programs that are better balanced, more dynamic, and more appropriate to the intellectual needs and abilities of children.

2. *The K–4 curriculum should actively involve children in doing mathematics.* Young children are active individuals who construct, modify, and integrate ideas by interacting with the physical world, materials, and other children. Given these facts, it is clear that the learning of mathematics must be an active process. Throughout the Standards, such verbs as *explore, justify, represent, solve, construct, discuss, use, investigate, describe, develop*, and *predict* are used to convey this active physical and mental involvement of children in learning the content of the curriculum.

The importance of active learning by children has many implications for mathematics education. Teachers need to create an environment that encourages children to explore, develop, test, discuss, and apply ideas. They need to listen carefully to children and to guide the development of their ideas. They need to make extensive and thoughtful use of physical materials to foster the learning of abstract ideas.

K–4 classrooms need to be equipped with a wide variety of physical materials and supplies. Classrooms should have ample quantities of such materials as counters; interlocking cubes; connecting links; base-ten, attribute, and pattern blocks; tiles; geometric models; rulers; spinners; colored rods; geoboards; balances; fraction pieces; and graph, grid, and dot paper. Simple household objects, such as buttons, dried beans, shells, egg cartons, and milk cartons, also can be used.

3. *The K–4 curriculum should emphasize the development of children's mathematical thinking and reasoning abilities.* An individual's future uses and needs for mathematics make the ability to think, reason, and solve problems a primary goal for the study of mathematics. Thus, the curriculum must take seriously the goal of instilling in students a sense of confidence in their ability to think and communicate mathematically, to solve problems, to demonstrate flexibility in working with mathematical ideas and problems, to make appropriate decisions in selecting strategies and techniques, to recognize familiar mathematical structures in unfamiliar settings, to detect patterns, and to analyze data. The K–4 standards reflect the view that mathematics instruction should promote these abilities so that students understand that knowledge is empowering and that individual pieces of content are all related to this broader perspective.

Developing these characteristics in children requires that schools build appropriate reasoning and problem-solving experiences into the curriculum from the outset. Further, this goal needs to influence the way mathematics is taught and the way students encounter and apply mathematics throughout their education.

4. *The K–4 curriculum should emphasize the application of mathematics.* If children are to view mathematics as a practical, useful subject, they must understand that it can be applied to a wide variety of real-world problems and phenomena. Even though most mathematical ideas in the K–4 curriculum arise *from* the everyday world, they must be regularly applied *to* real-world situations. Children also need to understand that mathematics is an integral part of real-world situations and activities in other curricular areas. The mathematical aspects of that work should be highlighted.

Learning mathematics has a purpose. At the K–4 level, one major purpose is helping children understand and interpret their world and solve problems that occur in it. Children learn computation to solve problems; they learn to measure because measurement helps them answer questions about how much, how big, how long, and so on; and they learn to collect and organize data because doing so permits them to answer other questions. By applying mathematics, they learn to appreciate the power of mathematics.

5. *The K–4 curriculum should include a broad range of content.* To become mathematically literate, students must know more than arithmetic. They must possess a knowledge of such important branches of mathematics as measurement, geometry, statistics, probability, and algebra. These increasingly important and useful branches of mathematics have significant and growing applications in many disciplines and occupations.

The curriculum at all levels needs to place substantial emphasis on these branches of mathematics. Mathematical ideas grow and expand as children work with them throughout the curriculum. The informal approach at this level establishes the foundation for further study and permits children to acquire additional knowledge they will need. These topics are highly appropriate for young learners because they make important contributions to children's mathematical development and help them see the usefulness of mathematics. They also provide productive, intriguing activities and applications.

The inclusion of a broad range of content in the curriculum also allows children to see the interrelated nature of mathematical knowledge. When teachers take advantage of the opportunity to relate one mathematical idea to others and to other areas of the curriculum, as will be described in Standard 4, children acquire broader notions about the interconnectedness of mathematics and its relationships to other fields. The curriculum should enable all children to do a substantial amount of work in each of these topics at each grade level.

6. *The K–4 curriculum should make appropriate and ongoing use of calculators and computers.* Calculators must be accepted at the K–4 level as valuable tools for learning mathematics. Calculators enable children to explore number ideas and patterns, to have valuable concept-development experiences, to focus on problem-solving processes, and to investigate realistic applications. The thoughtful use of calculators can increase the quality of the curriculum as well as the quality of children's learning.

Calculators do not replace the need to learn basic facts, to compute mentally, or to do reasonable paper-and-pencil computation. Classroom experience indicates that young children take a commonsense view about calculators and recognize the importance of not relying on them when it is more appropriate to compute in other ways. The availability of calculators means, however, that educators must develop a broader view of the various ways computation can be carried out and must place less emphasis on complex paper-and-pencil computation. Calculators also highlight the importance of teaching children to recognize whether computed results are reasonable.

The power of computers also needs to be used in contemporary mathematics programs. Computer languages that are geometric in nature help young children become familiar with important geometric ideas. Computer simulations of mathematical ideas, such as modeling the renaming of numbers, are an important aid in helping children identify the key features of the mathematics. Many software programs provide interesting problem-solving situations and applications.

The thoughtful and creative use of technology can greatly improve both the quality of the curriculum and the quality of children's learning. Integrating calculators and computers into school mathematics programs is critical in meeting the goals of a redefined curriculum.

SUMMARY OF CHANGES IN CONTENT AND EMPHASIS IN K–4 MATHEMATICS

INCREASED ATTENTION

NUMBER
- Number sense
- Place-value concepts
- Meaning of fractions and decimals
- Estimation of quantities

OPERATIONS AND COMPUTATION
- Meaning of operations
- Operation sense
- Mental computation
- Estimation and the reasonableness of answers
- Selection of an appropriate computational method
- Use of calculators for complex computation
- Thinking strategies for basic facts

GEOMETRY AND MEASUREMENT
- Properties of geometric figures
- Geometric relationships
- Spatial sense
- Process of measuring
- Concepts related to units of measurement
- Actual measuring
- Estimation of measurements
- Use of measurement and geometry ideas throughout the curriculum

PROBABILITY AND STATISTICS
- Collection and organization of data
- Exploration of chance

PATTERNS AND RELATIONSHIPS
- Pattern recognition and description
- Use of variables to express relationships

PROBLEM SOLVING
- Word problems with a variety of structures
- Use of everyday problems
- Applications
- Study of patterns and relationships
- Problem-solving strategies

INSTRUCTIONAL PRACTICES
- Use of manipulative materials
- Cooperative work
- Discussion of mathematics
- Questioning
- Justification of thinking
- Writing about mathematics
- Problem-solving approach to instruction
- Content integration
- Use of calculators and computers

♦ ♦ ♦ ♦ ♦ ♦ ♦ ♦

DECREASED ATTENTION

NUMBER
- Early attention to reading, writing, and ordering numbers symbolically

OPERATIONS AND COMPUTATION
- Complex paper-and-pencil computations
- Isolated treatment of paper-and-pencil computations
- Addition and subtraction without renaming
- Isolated treatment of division facts
- Long division
- Long division without remainders
- Paper-and-pencil fraction computation
- Use of rounding to estimate

GEOMETRY AND MEASUREMENT
- Primary focus on naming geometric figures
- Memorization of equivalencies between units of measurement

PROBLEM SOLVING
- Use of clue words to determine which operation to use

INSTRUCTIONAL PRACTICES
- Rote practice
- Rote memorization of rules
- One answer and one method
- Use of worksheets
- Written practice
- Teaching by telling

STANDARD 1: MATHEMATICS AS PROBLEM SOLVING

In grades K–4, the study of mathematics should emphasize problem solving so that students can—

- ***use problem-solving approaches to investigate and understand mathematical content;***
- ***formulate problems from everyday and mathematical situations;***
- ***develop and apply strategies to solve a wide variety of problems;***
- ***verify and interpret results with respect to the original problem;***
- ***acquire confidence in using mathematics meaningfully.***

Focus

Problem solving should be the central focus of the mathematics curriculum. As such, it is a primary goal of all mathematics instruction and an integral part of all mathematical activity. Problem solving is not a distinct topic but a process that should permeate the entire program and provide the context in which concepts and skills can be learned.

This standard emphasizes a comprehensive and rich approach to problem solving in a classroom climate that encourages and supports problem-solving efforts. Ideally, students should share their thinking and approaches with other students and with teachers, and they should learn several ways of representing problems and strategies for solving them. In addition, they should learn to value the process of solving problems as much as they value the solutions. Students should have many experiences in creating problems from real-world activities, from organized data, and from equations.

In the early years of the K–4 program, most problem situations will arise from school and other everyday experiences. When mathematics evolves naturally from problem situations that have meaning to children and are regularly related to their environment, it becomes relevant and helps children link their knowledge to many kinds of situations. As children progress through the grades, they should encounter more diverse and complex types of problems that arise from both real-world and mathematical contexts.

When problem solving becomes an integral part of classroom instruction and children experience success in solving problems, they gain confidence in doing mathematics and develop persevering and inquiring minds. They also grow in their ability to communicate mathematically and use higher-level thinking processes.

Discussion

Classrooms with a problem-solving orientation are permeated by thought-provoking questions, speculations, investigations, and explorations; in this environment, the teacher's primary goal is to promote a problem-solving approach to the learning of all mathematics content. The following two examples illustrate this meaning.

A lesson designed to develop the characteristics of parallelograms can

Fig. 1.1

be approached from a problem-solving perspective. The teacher, who has a collection of quadrilaterals like the ones shown in figure 1.1, has the children discover the teacher's rule for sorting the shapes. One rule is to have all parallelograms in one loop and all nonparallelograms in the other.

In turn, each child picks a shape and decides in which loop to put it. The teacher says yes or no as each student places a shape. Throughout the process, the children are asked to think about the common characteristics of the shapes in each loop; after all shapes are placed, these common characteristics are discussed. As a result, the children learn to define parallelograms and name their characteristics in the context of a thought-provoking activity.

Basic subtraction facts also can be presented in a problem-solving setting.

Each child is asked to put 13 small counters under one hand and, without looking, to move 6 of them into view. The teacher asks, "Can you figure out how many counters are still under your hand?" The children are invited to share their solution strategies. Responses might include the following:

There are six over here [outside]; six more would be twelve, so there must be seven left.

You have six here. Four more make ten and three more make thirteen; four and three are seven. Seven are left.

The class then discusses solving subtraction problems by "adding on."

Once again, the mathematical ideas have originated with the children rather than the teacher, in an inquiry-oriented manner.

Computer software also is a significant component of a comprehensive problem-solving program. Many excellent software packages enable children to develop and apply problem-solving strategies in geometry, logical reasoning, classification, measurement, fractions and decimals, and other mathematical content.

A major goal of problem-solving instruction is to enable children to develop and apply strategies to solve problems. Strategies include using manipulative materials, using trial and error, making an organized list or table, drawing a diagram, looking for a pattern, and acting out a problem.

Consider the following problem:

I have some pennies, nickels, and dimes in my pocket. I put three of the coins in my hand. How much money do you think I have in my hand?

pennies	nickels	dimes	total value
0	0	3	30
0	1	2	25
0	2	1	20
0	3	0	15
1	0	2	21
⋮	⋮	⋮	⋮

Fig. 1.2

This problem leads children to adopt a trial-and-error strategy. They can also act out the problem by using real coins. Children verify that their answers meet the problem conditions. Follow-up questions can also be posed: "Is it possible for me to have four cents? Eleven cents? Can you list all the possible amounts I can have when I pick three coins?" The last question provides a challenge for older or more mathematically sophisticated children and requires them to make an organized list of possible coin combinations, perhaps like the one in figure 1.2.

The initial conditions can be altered to include quarters:

I have six coins worth 42 cents; what coins do you think I have? Is there more than one answer?

A vital component of problem-solving instruction is having children formulate problems themselves. Children can write variations for problems previously explored, word problems that correspond to a number sentence, or a question that can be answered by investigating data in a menu, advertisement, or chart, like the one in figure 1.3, which lists children's eye colors. A primary feature of this context is that it lends itself to the use of calculators. Using calculators in problem-solving settings to perform tedious calculations enables children to focus on the problem-solving processes rather than on the calculations.

EYE COLOR

	Grade 1			Grade 2			Grade 3		
	Blue	Brown	Other	Blue	Brown	Other	Blue	Brown	Other
School 1	346	219	24	304	206	83	381	162	47
School 2	328	46	23	289	53	42	341	22	37

Fig. 1.3

Children might pose such questions as, "How many children are in each school? How many more blue-eyed children than brown-eyed students are in grade 1? In all? Why do you think school 2 is different from school 1?"

Project problems, which often require several days of class time, provide an opportunity for children to become immersed in problem-solving activity. Some situations allow children to be particularly creative in their formulation of problems. Here is one such situation:

The class is given the opportunity to plan and participate in an all-school "Estimation Day." The children, in pairs or threes, are to design estimation activities to be completed by children in other classes. Each group will supply all the necessary materials and monitor the activities. The activities might include guessing children's heights, the number of candies in a jar, the lengths of various pieces of string, the weight of a bag of potatoes, the length of the room, the number of times they can write their names in a minute, or the length of time required for an ice cube to melt.

Participation in project problems allows children to acquire confidence in their problem-solving ability. In working with others, they become a vital part of a team and find that their contributions are essential to the success of the project.

When problem solving is an integral part of the curriculum, beginning with a child's earliest encounters with mathematics, children develop a point of view about what it means to learn mathematics and solve problems in mathematics.

STANDARD 2: MATHEMATICS AS COMMUNICATION

In grades K–4, the study of mathematics should include numerous opportunities for communication so that students can—

- ***relate physical materials, pictures, and diagrams to mathematical ideas;***
- ***reflect on and clarify their thinking about mathematical ideas and situations;***
- ***relate their everyday language to mathematical language and symbols;***
- ***realize that representing, discussing, reading, writing, and listening to mathematics are a vital part of learning and using mathematics.***

Focus

Mathematics can be thought of as a language that must be meaningful if students are to communicate mathematically and apply mathematics productively. Communication plays an important role in helping children construct links between their informal, intuitive notions and the abstract language and symbolism of mathematics; it also plays a key role in helping children make important connections among physical, pictorial, graphic, symbolic, verbal, and mental representations of mathematical ideas. When children see that one representation, such as an equation, can describe many situations, they begin to understand the power of mathematics; when they realize that some ways of representing a problem are more helpful than others, they begin to understand the flexibility and usefulness of mathematics.

Young children learn language through verbal communication; it is important, therefore, to provide opportunities for them to "talk mathematics." Interacting with classmates helps children construct knowledge, learn other ways to think about ideas, and clarify their own thinking. Writing about mathematics, such as describing how a problem was solved, also helps students clarify their thinking and develop deeper understanding. Reading children's literature about mathematics, and eventually text material, also is an important aspect of communication that needs more emphasis in the K–4 curriculum.

This standard highlights the need to involve children in actively doing mathematics. Exploring, investigating, describing, and explaining mathematical ideas promote communication. Teachers facilitate this process when they pose probing questions and invite children to explain their thinking. Teachers also can assess students' knowledge and insight by listening and observing. The idea that children should learn mathematics meaningfully is implicit in this discussion, for meaningful learning is necessary if mathematics is to make sense and if communication is to be possible.

Discussion

Young children are active, social individuals. Much of the sense they make of the world is derived from their communications with other people. Communicating helps children to clarify their thinking and sharpen their understandings.

♦ ♦ ♦ ♦ ♦ ♦ ♦ ♦

Representing, talking, listening, writing, and reading are key communication skills and should be viewed as integral parts of the mathematics curriculum. Probing questions that encourage children to think and explain their thinking orally or in writing help them to understand more clearly the ideas they are expressing.

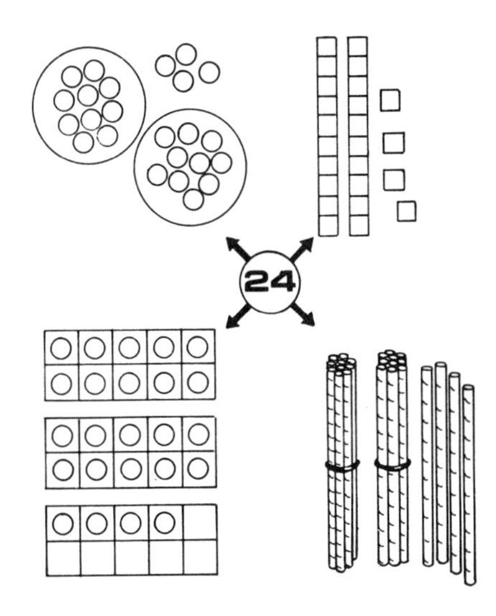

Fig. 2.1

Representing is an important way of communicating mathematical ideas at all levels, but especially so in K–4. Representing involves translating a problem or an idea into a new form. Translations of this type often are used by adults and children as they converse with others. Children might draw diagrams, for example, to express an idea or viewpoint in an alternative format that might be more comprehensible to the listener. The act of representing encourages children to focus on the essential characteristics of a situation. Representing includes the translation of a diagram or physical model into symbols or words. A child should be able to examine a set of two bundles of ten and four units each and match the set with the symbol 24. (See fig. 2.1.) Since representing is central to learning and using mathematics, it is important to provide many such experiences for children. Any of the following materials would be useful: base-ten blocks, straws that can be bundled in sets of ten, connecting cubes, or loose counters that must simply be grouped together to show the tens.

Representing is also used in translating or analyzing a verbal problem to make its meaning clear. The problem in figure 2.2 involves a part-whole comparison, but many children who have not had enough experience in modeling such situations or who do not actively model this situation fail to recognize its structure.

If a bag contains 3 white balls and 1 blue ball, what is the probability of pulling out a blue ball?

Fig. 2.2

Some students simply use the numbers 1 and 3 to arrive at an answer of 1/3. A concrete representation or diagram will help children to correctly identify that four balls can possibly be drawn.

Communicating by talking and listening is also very important. When small groups of children discuss and solve problems, they are able to connect the language they know with mathematical terms that might be unfamiliar to them. They make sense of those problems. The use of concrete materials is particularly appropriate because they give the children an initial basis for conversation. Such occasions also permit the teacher to observe individual students, to ask probing questions, and to note or attend to any conceptual difficulties individual students might be experiencing. The following discussion activity would help children see how several problems that appear to be different in fact share the same underlying structure: $14 - 5 = \square$. The children would be given counters to model each problem.

With your group, use counters of different colors to model each of these problems and then discuss how the problems are alike or different.

Maria had some pencils in her desk. She put 5 more in her desk. Then she had 14. How many pencils did she have in her desk to start with?

Eddie had 14 helium balloons. Several of them floated away. He had 5 left. How many did he lose?

Nina had 14 seashells. That was 5 more than Pedro had. How many seashells did Pedro have?

$14 - 5 = \square$

As children talk about mathematics, it is important to keep in mind that what appears at first to be an incorrect response may be, in fact, an inability to communicate. Of primary importance is the value children derive from reflecting on their responses.

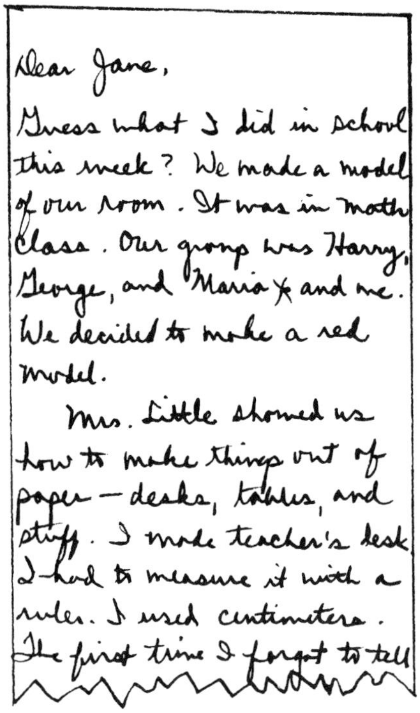

Dear Jane,

Guess what I did in school this week? We made a model of our room. It was in math class. Our group was Harry, George, and Maria and me. We decided to make a red model.

Mrs. Little showed us how to make things out of paper—desks, tables, and stuff. I made teacher's desk. I had to measure it with a ruler. I used centimeters. The first time I forgot to tell

Fig. 2.3

Writing is a communication skill that has been used too infrequently in mathematics. It is particularly useful because it allows a child who is uncomfortable in oral situations to express understanding in a less public forum. After children have solved a problem, they can write their answer in sentence form, which helps them exhibit a knowledge of the problem's place in the real world and clarify their thinking.

Students can write a letter to tell a friend about something they have learned in mathematics class. This type of activity allows the students to consider mathematics for a new purpose. If letters are exchanged, then students learn from the thought processes of their peers. See figure 2.3.

Having students keep journals in mathematics class is another way to facilitate communication and give them an opportunity to reflect on their learning. A journal can be a form of free expression about the mathematics studied, or children can be asked to respond to directions such as these: Tell me what you thought were the hardest and easiest parts of today's lesson and why.

Children can also create their own stories or books about mathematics. Many schools have a "young authors" program that encourages children to develop an idea into a book to be shared with parents or classmates. This activity is within the reach of fourth graders and can include mathematics topics as options for development.

Many children's books present interesting problems and illustrate how other children solve them. Through these books students see mathematics in a different context while they use reading as a form of communication. Some of the books most directly linked to mathematics give children insights into the history of mathematics and the development of mathematical ideas. Materials children write themselves can be part of a reading activity and shared with class members. Mathematics texts have not often been viewed as sources of reading material, but taking this perspective can add a valuable dimension to students' learning. Many schools are making efforts to include expository reading as an important part of reading instruction. Mathematics texts and other mathematics reading materials should certainly be included in these efforts.

Children learn from one another as they communicate. Encouraging them to represent, talk and listen, write, and read facilitates meaningful learning. Attending to students' communications about their thinking also gives teachers a rich information base from which they can make sound instructional decisions.

STANDARD 3: MATHEMATICS AS REASONING

In grades K–4, the study of mathematics should emphasize reasoning so that students can—

- ***draw logical conclusions about mathematics;***
- ***use models, known facts, properties, and relationships to explain their thinking;***
- ***justify their answers and solution processes;***
- ***use patterns and relationships to analyze mathematical situations;***
- ***believe that mathematics makes sense.***

Focus

A major goal of mathematics instruction is to help children develop the belief that they have the power to do mathematics and that they have control over their own success or failure. This autonomy develops as children gain confidence in their ability to reason and justify their thinking. It grows as children learn that mathematics is not simply memorizing rules and procedures but that mathematics makes sense, is logical, and is enjoyable. A classroom that values reasoning also values communicating and problem solving, all of which are components of the broad goals of the entire elementary school curriculum.

A climate should be established in the classroom that places critical thinking at the heart of instruction. Both teachers' and children's statements should be open to question, reaction, and elaboration from others in the classroom. Such a climate depends on all members of the class expressing genuine respect and support for one another's ideas. Children need to know that being able to explain and justify their thinking is important and that how a problem is solved is as important as its answer. This mind-set is established when children have opportunities to apply their reasoning skills and when justifying one's thinking is an expected component of problem discussions.

Discussion

This standard's descriptor, "Mathematics as Reasoning," was purposely chosen. Mathematics *is* reasoning. One cannot do mathematics without reasoning. The standard does not suggest, however, that formal reasoning strategies be taught in grades K–4. At this level, mathematical reasoning should involve the kind of informal thinking, conjecturing, and validating that helps children to see that mathematics makes sense. Consistent use of such questions as "Why do you think that's a good answer?" or "Do you think that you would get the same answer if you used these other materials?" conveys to the children the importance of critical thinking and establishes a spirit of inquiry.

Children should be encouraged to justify their solutions, thinking processes, and conjectures in a variety of ways. Manipulatives and other physical models help children relate processes to their conceptual underpinnings and give them concrete objects to talk about in explaining and justifying their thinking. Observing children interact with objects in this way allows teachers to reinforce thinking processes and evaluate any possible misunderstandings.

Creating and extending patterns of manipulative materials and recognizing relationships within patterns require children to apply analytical and spatial reasoning. See figure 3.1.

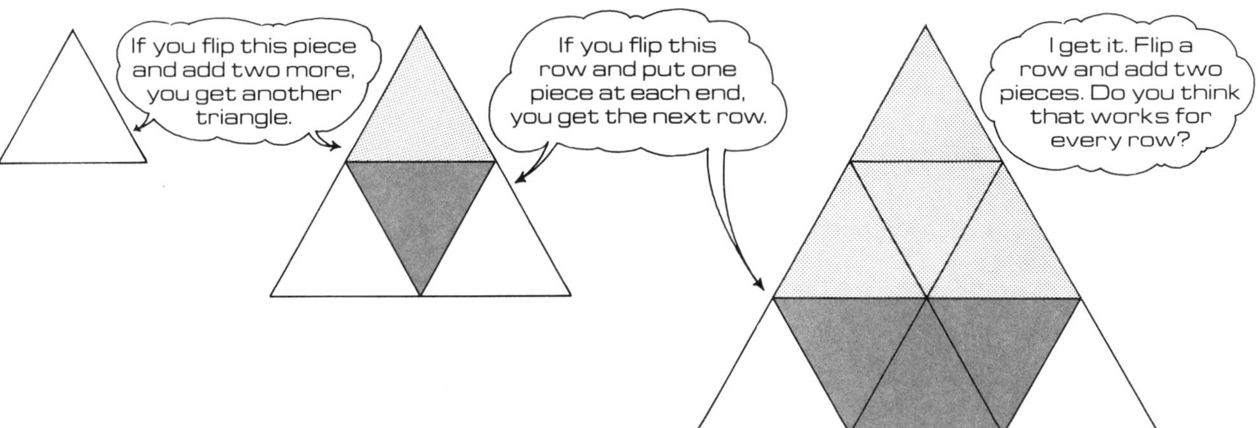

Fig. 3.1

The kindergartner who created the attribute block pattern in figure 3.2 proudly announced that she had four patterns in one.

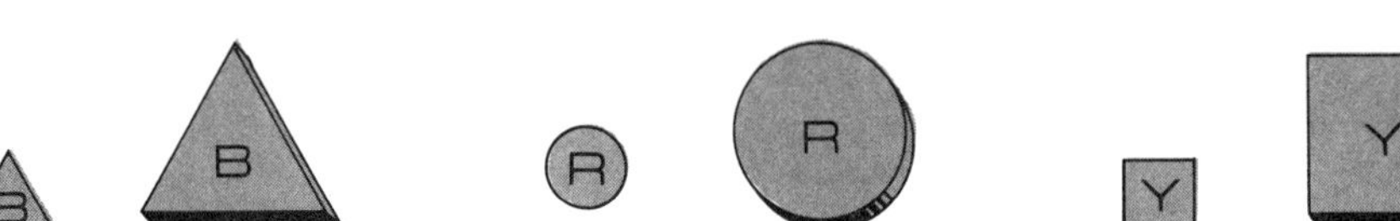

Fig. 3.2

Pointing to each element in turn, she said, "See, there's triangle, triangle, circle, circle, square, and square. That's one pattern. Then there's small, large, small, large, small, and large. That's the second pattern. Then there's thin, thick, thin, thick, thin, and thick. That's the third pattern. And the fourth pattern is blue, blue, red, red, yellow, and yellow. The triangles are blue, the circles are red, and the squares are yellow."

Children also reason analytically when they identify valid arguments. When the class considers 35 − 19 = □, the teacher can ask questions like these: "Do you think it would help to know that 35 − 20 = 15?" "How would it help to think of 19 as 15 + 4?" "Would it help to count on from 19 to 35?" It is also important for children to recognize invalid arguments, such as "Would it help to count backward from 19?"

Many problems can be solved by the process of elimination, in which children systematically select the items that satisfy one or more given conditions by eliminating those that do not. "Who Am I?" and "What Am I?" games require this kind of thinking in both creating and solving problems. See figure 3.3.

What Am I?

I have 3 or 4 sides.
All my angles are equal.
My sides are not all equal.

Who Am I?

I am an even number.
I am more than 20 and less than 30.
I am not 25.
The sum of my digits is 8.

Fig. 3.3

These activities also give students a chance to encounter informally several important ideas, such as the language of logic, the use of a counterexample, and distinctions between relevant and irrelevant information. The use of *and, or,* and *not* in these activities illustrates the language of logic. The reasoning in the "What Am I?" game is based on a counterexample involving known properties of an equilateral triangle. "If I have three sides, I am a triangle. But I can't be a triangle because all triangles with equal angles have equal sides and I do not have equal sides." The clue "I am not 25" in the "Who Am I?" game is irrelevant because another clue identifies the number as even: Clearly the number cannot be 25, and this information is of no value in solving the problem.

Applying reasoning skills to discover a relationship they have not recognized before can be an exhilarating experience for children, as a group of third graders learned. They were using a calculator to explore number relationships when they noticed that if one addend is decreased by any amount and another addend is increased by the same amount, their sum remains the same. After checking their conjecture with a variety of numbers, they recorded it as a discovery so that it could be shared with the rest of the class. See figure 3.4

19	18	16	13
+5	+6	+8	+11
24	24	24	24

123	120	100
+76	+79	+99
199	199	199

Fig. 3.4

Our Discovery: When you add, if you make one part bigger and the other part gets the same amount smaller, you always get the same answer.

One member of the group thought the relationship should "work" for subtraction, too, until a partner showed several cases for which it did not work.

These children applied analytical reasoning and developed and tested conjectures, one of which they rejected on the basis of counterexamples.

An informal introduction to proportional reasoning is appropriate at the K–4 level. The problem-solving context of the following example also reinforces many of the reasoning processes already discussed.

I have a shape that can be covered with twelve of these triangles. How many of these parallelograms would I need to cover my shape? How many of these trapezoids will cover my shape?

Since the problem concerns physical objects, the students can recognize visually that two triangles cover a parallelogram, that three triangles cover a trapezoid, and that the entire shape can be covered by six parallelograms or by four trapezoids. To justify their conclusion, the children can use twelve triangles to make a shape and check to see whether it can be covered by six parallelograms or four trapezoids. Students should also realize that some shapes composed of twelve triangles cannot be covered by parallelograms or trapezoids.

Mathematical reasoning cannot develop in isolation. As illustrated in this discussion, the ability to reason is a process that grows out of many experiences that convince children that mathematics makes sense.

STANDARD 4: MATHEMATICAL CONNECTIONS

In grades K–4, the study of mathematics should include opportunities to make connections so that students can—

- ***link conceptual and procedural knowledge;***
- ***relate various representations of concepts or procedures to one another;***
- ***recognize relationships among different topics in mathematics;***
- ***use mathematics in other curriculum areas;***
- ***use mathematics in their daily lives.***

Focus

This standard's purpose is to help children see how mathematical ideas are related. The mathematics curriculum is generally viewed as consisting of several discrete strands. As a result, computation, geometry, measurement, and problem solving tend to be taught in isolation. It is important that children connect ideas both among and within areas of mathematics. Without such connections, children must learn and remember too many isolated concepts and skills rather than recognizing general principles relevant to several areas. When mathematical ideas are also connected to everyday experiences, both in and out of school, children become aware of the usefulness of mathematics.

A classroom in which making connections is emphasized exhibits several notable characteristics. Ideas flow naturally from one lesson to another, rather than each lesson being restricted to a narrow objective. Lessons frequently extend over several days so that connections can be explored, discussed, and generalized. Once introduced, a topic is used throughout the mathematics program. Teachers seize opportunities that arise from classroom situations to relate different areas and uses of mathematics. Children are asked to compare and contrast concepts and procedures. They are helped to construct bridges between the concrete and the abstract and between different ways of representing a problem or concept. Learning and using mathematics are important aspects of the entire school curriculum.

Discussion

When children enter school, they have not segregated their learning into separate school subjects or topics within an academic area. Thus, it is particularly important to build on the wholeness of their perspective of the world and expand it to include more of the world of mathematics. This can be done in many ways, both within and outside the realm of mathematics.

Young children understand the underlying structure of many numerical problems and use counting to solve them. It is important to tie these conceptual ideas to more abstract procedures such as adding and subtracting. If conceptual understandings are linked to procedures, children will not perceive of mathematics as an arbitrary set of rules; will not need to learn or memorize as many procedures; and will have the foundation to apply, re-create, and invent new ones when needed. For example,

◆ ◆ ◆ ◆ ◆ ◆ ◆ ◆

if children are asked to fold paper and describe the process, they will understand why the procedure "multiplying the numerator and denominator by the same number" yields the same ratio in an equivalent fraction. See figure 4.1.

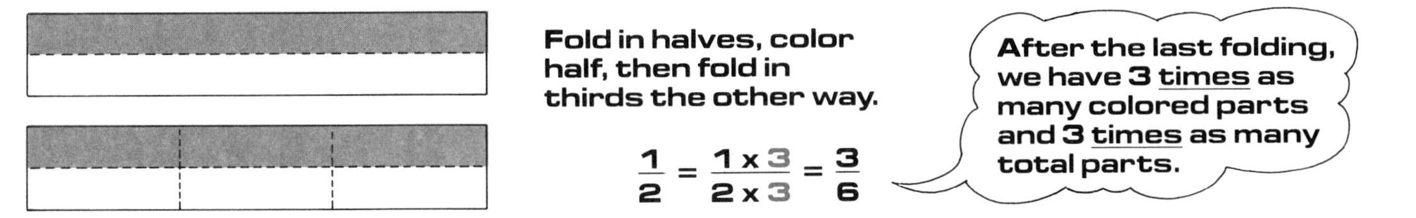

Fig. 4.1

Many concrete and pictorial models of concepts and procedures are available, and children need to create relationships among them and determine how each can be represented with symbols. For example, young children need to make the connection between seven toy cars, seven counters, seven tally marks, and the symbol 7. Older children need to understand the similarity between cutting a rectangle into four equal parts and sharing a bag of cookies among four friends and why the parts in each situation are called fourths. They need to see different representations of the same problem situation, as in figure 4.2.

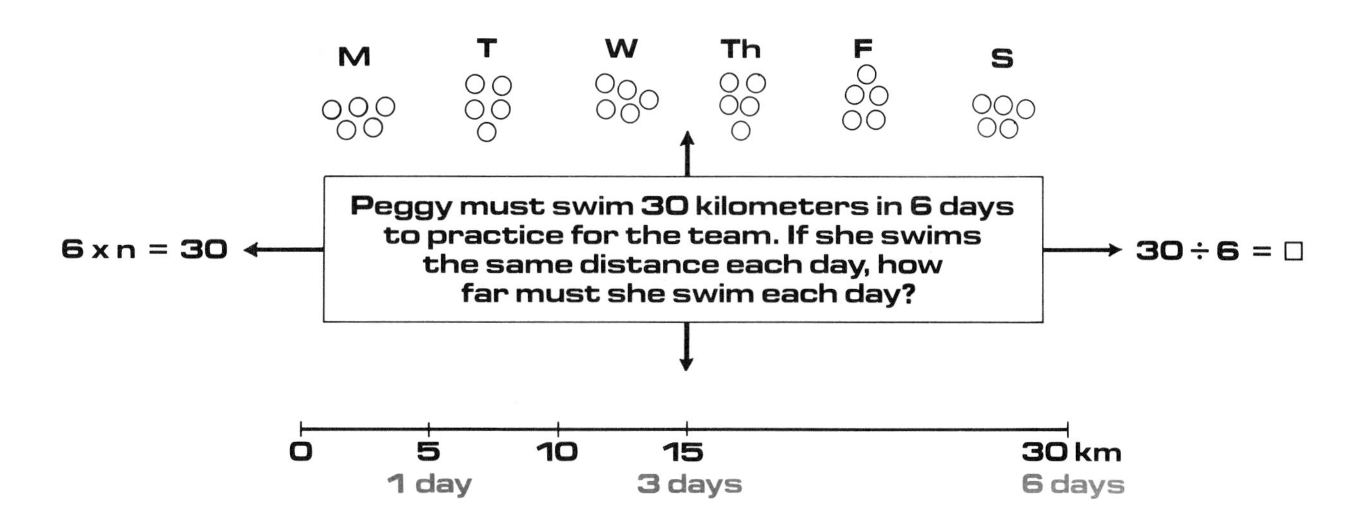

Fig. 4.2

Children tend to think of mathematics as computation. One way to dispel this incorrect notion is to offer them more experiences with other topics; even so, unless connections are made, children will see mathematics as a collection of isolated topics. Only through extended exposure to integrated topics will children have a better chance of retaining the concepts and skills they are taught. For example, measurement situations should continually be part of the program, rather than introduced briefly in isolated lessons. The following activity integrates geometry with measurement.

12 cm

16 cm

Cut a 12-by-16-cm rectangle on a diagonal as shown. What geometric shapes can you make? Which one has the shortest perimeter?

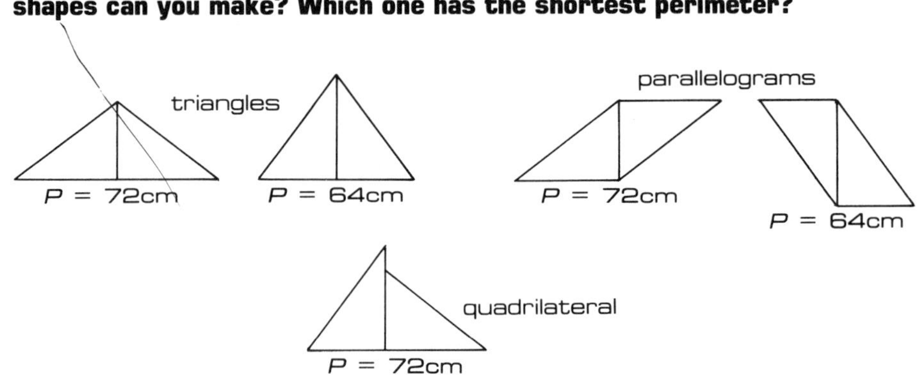

Similarly, addition practice can be placed in the context of measuring as children solve for the distances between cages at a zoo (see fig. 4.3).

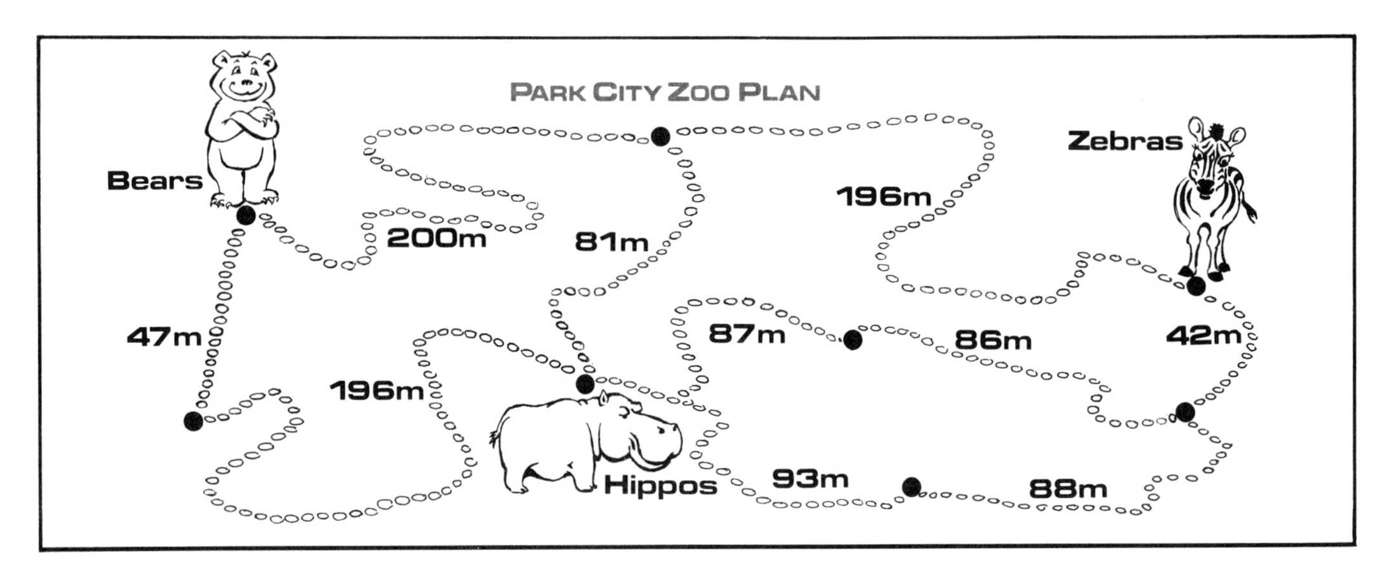

Find the shortest distance from the:

bears to the hippos ____

zebras to the bears ____

hippos to the zebras ____

zebras to the hippos ____

Fig. 4.3

Another connection children can explore is that between solutions to open number sentences and graphing, as shown below.

$\square + \square + \Delta = 9$

Try $\square = 1$

$1 + 1 + \Delta = 9$

Δ must be 7

$2 + 2 + 5 = 9$
$3 + 3 + 3 = 9$
$4 + 4 + 1 = 9$
$0 + 0 + 9 = 9$

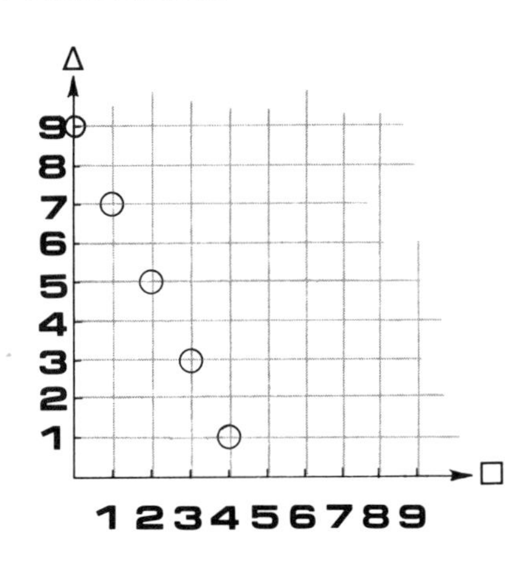

The K–4 program is rich with opportunities to use mathematics in other subject areas as well as to use other subjects in mathematics. This is especially true with science, but with a little imagination connections can be made to all areas. For example, the communication standard (Standard 2) calls for the integration of language arts as children write and discuss their experiences in mathematics. As children solve problems in mathematics classes, they can be learning about other countries and cultures. As children measure how far they can jump, they use mathematics in physical education. As children do art projects, they use geometry and measurement.

Social Studies

What is the tallest building in Japan?
How tall is it?
Write a sentence comparing its height to the height of the Sears Tower in Chicago.

Physical Education

Measure how far you can go in three jumps.

One frog jump.
Start and end on hands and feet.

Art

Make the following picture frame for your miniature drawing.

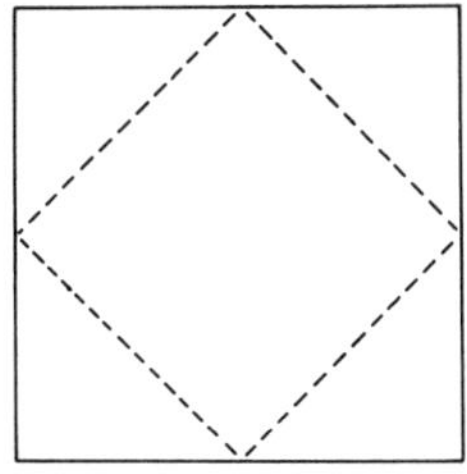

Cut a 20-cm square.
Find midpoints of the sides.

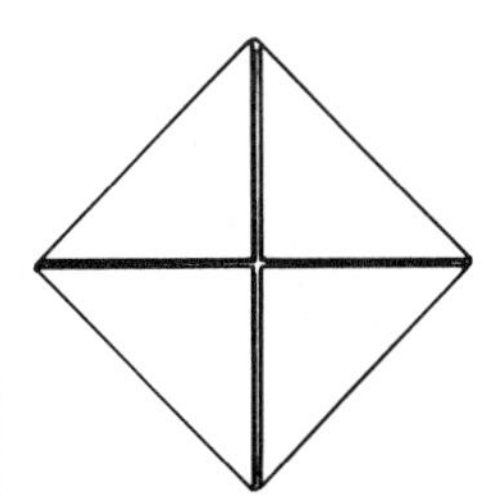

Fold flaps back.
Find midpoints of the sides.

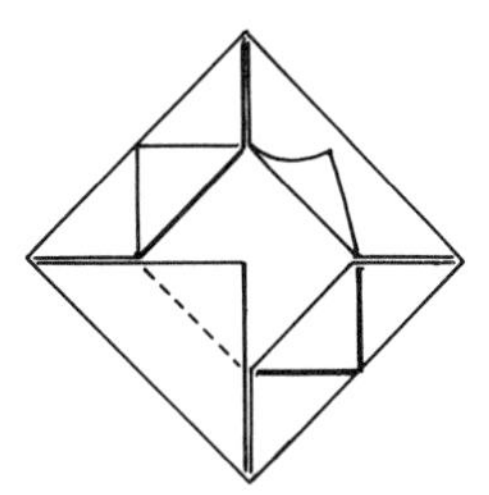

Fold flaps back to midpoints.

Fold flaps under and glue picture in frame.

All too often, children come to believe that mathematics is an academic exercise that occurs only in schools, whereas solving problems outside of school is different. Many believe that it is not mathematics to explore the meaning of one-third by sharing a pitcher of milk equally among three people; to count on a clock face how long until it is time to go to a friend's house; or to figure 100 ÷ 4 by thinking, "Four quarters make a dollar, so it's 25," or "It's 100 divided by 2 and 2 again." Mathematical methods exist to solve these problems in an efficient manner, but, at times, these are not as satisfactory as the informal ways. Students need to see when and how mathematics can be used, rather than be promised that someday they will use it.

STANDARD 5: ESTIMATION

In grades K–4, the curriculum should include estimation so that students can—

- ***explore estimation strategies;***
- ***recognize when an estimate is appropriate;***
- ***determine the reasonableness of results;***
- ***apply estimation in working with quantities, measurement, computation, and problem solving.***

Focus

Estimation presents students with another dimension of mathematics; terms such as *about, near, closer to, between,* and *a little less than* illustrate that mathematics involves more than exactness. Estimation interacts with number sense and spatial sense to help children develop insights into concepts and procedures, flexibility in working with numbers and measurements, and an awareness of reasonable results. Estimation skills and understanding enhance the abilities of children to deal with everyday quantitative situations.

From children's earliest experiences with mathematics, estimation needs to be an ongoing part of their study of numbers, computation, and measurement. It is important that children learn a variety of methods of estimating, such as the front-end strategy for computation and the chunking procedure for measurement. They also need to develop reasoning, judgment, and decision-making skills in using estimation.

Instruction should emphasize the development of an estimation mind-set. Children should come to know what is meant by an estimate, when it is appropriate to estimate, and how close an estimate is required in a given situation. If children are encouraged to estimate, they will accept estimation as a legitimate part of mathematics.

Discussion

When children enter school, they are accustomed to estimating. They know that they are almost six years old, that they are a little shorter than a brother or sister, that a carton of milk can fill more than three glasses, and when it is about noon. This experiential knowledge provides a foundation for further development in estimating quantities. Consider the following example: As a referent, children can be told that the set at the left in figure 5.1 has ten balls. Without counting, they can be asked to quickly classify the other sets as fewer than ten, about ten, or more than ten.

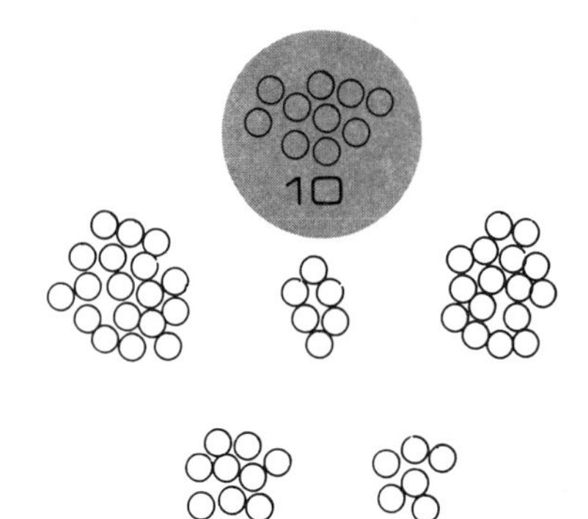

Fig. 5.1

Children should also estimate larger quantities, such as the number of seeds in a pumpkin, beans in a bag, or Valentine candies in a jar. For larger quantities, it is usually more appropriate to use a referent set having 50 or 100 items.

Several important considerations in estimating quantities should be remembered. When checking estimates, a teacher can reinforce place-value ideas by having the children place the estimated items in groups of ten and then in hundreds whenever possible. It is also important for the teacher and the children to identify a range for "good estimates." Further, it should be emphasized that estimates that happen to be exact are no better than other estimates within the identified range; the goal is an approximation, not the exact number. Finally, children should always check their initial estimates and then make additional ones so that they can use the feedback to refine their estimating skills.

A particularly good estimating activity involving measuring uses interlocking cubes. Using a stick of ten cubes as a referent, children estimate how many cubes long a work table is, for example. Then they make a "train" of cubes as long as the work table and break the train into sticks of ten to check their estimates. See figure 5.2.

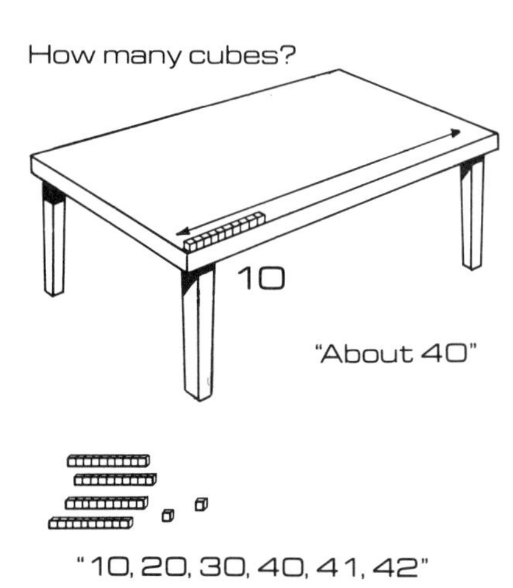

Fig. 5.2

Another measurement-and-estimating activity illustrates the process of chunking. In the task in figure 5.3, children estimate the number of boxes necessary to fill the classroom. A child mentally lines up seven boxes along one edge of the floor and uses them as a unit, or a "chunk," to estimate the total number of boxes.

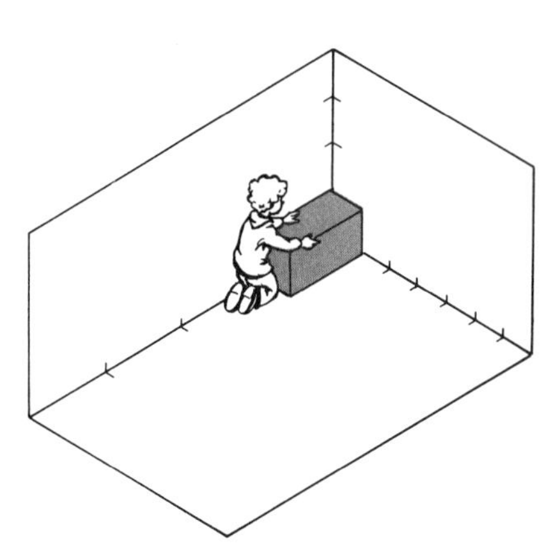

"I think seven boxes will fit along the floor; then I can do that three more times to cover the floor, that's 4 × 7 = 28, then four layers in all, that's 4 × 28 = 112."

Fig. 5.3

Children also should be taught specific strategies to aid them in *computational* estimation. A child who needs to evaluate 243 + 479 might estimate by thinking, "200 and 400 is 600, 43 and 79 is over 100, so the sum is a little more than 700." This is "front-end estimation." Another way of estimating is this: "243 is just under 250, 479 is just under 500, so the sum is less than 750." This is a flexible use of "rounding" for estimation or selecting "nice" numbers that are easy to work with. It is useful to discuss various strategies and to help students develop their own strategies. For example, a student adept at mental computation could estimate 243 + 479 in this way: "24 + 48 (tens) is 72 (tens) so the sum is about 720." Continual emphasis on computational estimation helps children develop creative and flexible thought processes and fosters in them a sense of mathematical power.

Estimation is especially important when children use calculators. If they need to compute 4783 ÷ 13, for example, a quick estimate can be found by using "compatible numbers." In this case, 4783 is about 4800 and 13 is about 12, so 4783 ÷ 13 is about 4800 ÷ 12. The dividing can be done mentally, since 48 and 12 are "compatible numbers" for division. Thus, 4783 ÷ 13 is about 400. This rough estimate provides children with enough information to decide whether the correct keys were pressed and whether the calculator result is reasonable. Such uses of estimation reduce the incidence of errors with calculators, decrease the mindless use of calculators for computation, and contribute to children's development of number and operation sense.

Children often find that estimation skills are useful in their daily lives. Many children know when it is appropriate to estimate and how close an estimate should be, as the following anecdote indicates. Three children huddled together in a shopping mall, discussing the purchase of some clothing. One held a newspaper advertisement, another a calculator. Two children picked items from the ad and the third entered the appropriate prices into the calculator. In considering the calculator result, one of the children reasoned, "The total cost can't be more than $50 because two shirts cost $14 each; that's less than $30, and the pants cost $17.99." Classroom instruction on estimation should help children develop a similar estimation mind-set so they can use good judgment and logical reasoning to make decisions in their daily lives.

STANDARD 6: NUMBER SENSE AND NUMERATION

In grades K–4, the mathematics curriculum should include whole number concepts and skills so that students can—

- ***construct number meanings through real-world experiences and the use of physical materials;***
- ***understand our numeration system by relating counting, grouping, and place-value concepts;***
- ***develop number sense;***
- ***interpret the multiple uses of numbers encountered in the real world.***

Focus

Children must understand numbers if they are to make sense of the ways numbers are used in their everyday world. They need to use numbers to quantify, to identify location, to identify a specific object in a collection, to name, and to measure. Furthermore, an understanding of place value is crucial for later work with number and computation.

Intuition about number relationships helps children make judgments about the reasonableness of computational results and of proposed solutions to numerical problems. Such intuition requires good number sense. Children with good number sense (1) have well-understood number meanings, (2) have developed multiple relationships among numbers, (3) recognize the relative magnitudes of numbers, (4) know the relative effect of operating on numbers, and (5) develop referents for measures of common objects and situations in their environments.

Children come to understand number meanings gradually. To encourage these understandings, teachers can offer classroom experiences in which students first manipulate physical objects and then use their own language to explain their thinking. This active involvement in, and expression of, physical manipulations encourages children to reflect on their actions and to construct their own number meanings. In all situations, work with number symbols should be meaningfully linked to concrete materials. Emphasizing exploratory experiences with numbers that capitalize on the natural insights of children enhances their sense of mathematical competency, enables them to build and extend number relationships, and helps them to develop a link between their world and the world of mathematics.

If children are to develop good number concepts, considerable instructional time must be devoted to number and numeration. Children's experiences with numbers are most beneficial when the numbers have meaning for them. A variety of place-value tasks that assess children's thinking can be used to identify those numbers that have meaning to individual students; traditional numeration tasks are not good indicators of children's understanding. Teachers can also provide exploratory experiences with larger numbers, but symbolic tasks with numbers should not be presented in isolation and should not be emphasized until the numerals have been carefully linked to concrete materials and children understand the major concepts.

◆ ◆ ◆ ◆ ◆ ◆ ◆ ◆

Discussion

For children to use both single-digit and multidigit number ideas fluently, written symbols should be linked to physical models and oral names. See figure 6.1.

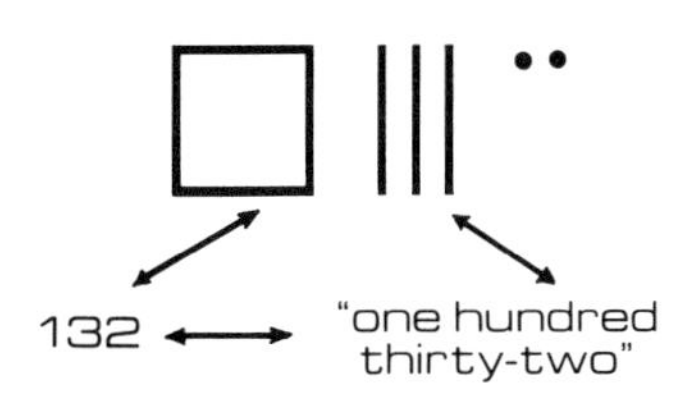

Fig. 6.1

Counting skills, which are essential for ordering and comparing numbers, are an important component of the development of number ideas. Counting on, counting back, and skip counting mark advances in children's development of number ideas. However, counting is only one indicator of children's understanding of numbers.

Understanding place value is another critical step in the development of children's comprehension of number concepts. Prior to formal instruction on place value, the meanings children have for larger numbers are typically based on counting by ones and the "one more than" relationship between consecutive numbers. Since place-value meanings grow out of grouping experiences, counting knowledge should be integrated with meanings based on grouping. Children are then able to use and make sense of procedures for comparing, ordering, rounding, and operating with larger numbers.

The following activity encourages children to coordinate their counting and grouping skills to develop beginning place-value ideas. Two children each are given the same number of counters, in this example, thirty-two. One child counts her counters by ones; the other groups his counters by tens and then counts by tens and ones. The children then are asked to compare and discuss their results.

"one, two, . . . thirty-two"

"ten, twenty, thirty, thirty-one, thirty-two"

The next two tasks help determine a child's place-value knowledge.

"Count these loose chips . . . [25]. Could you write that?" [25] The teacher circles the digit 5 and asks, "Does this part of your 25 have anything to do with how many chips you have?" She repeats the action, this time circling the digit 2. Children with good place-value knowledge will match the "5" with five chips and the "2" with twenty chips, and they may even group the twenty chips into two groups of ten chips. [Fig. 6.2]

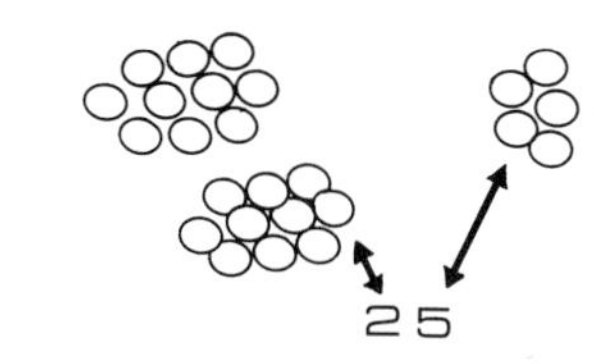

Fig. 6.2

"Here are 256 beans. How many piles of 10 beans could you make?" [Fig. 6.3]

Fig. 6.3

Number sense is an intuition about numbers that is drawn from all the varied meanings of number. It has five components:

1. *Developing number meanings.* This includes the cardinal and ordinal meanings of numbers.

2. *Exploring number relationships with manipulatives.* For example, the composition and decomposition of sets of objects enables children to understand 7 as shown in figure 6.4. Similarly, they understand that 50 is 5 tens, 2 twenty-fives, or 4 tens and 10 ones.

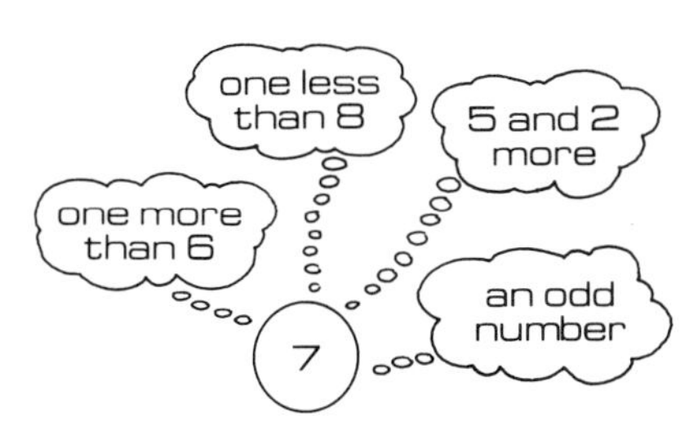

Fig. 6.4

3. *Understanding the relative magnitudes of numbers.* For example, 31 is large compared to 4, about the same size as 27, about half as big as 60, and small compared to 92. Counting by ones rapidly to 100 or 1000 on a calculator helps establish the relative sizes of these numbers.

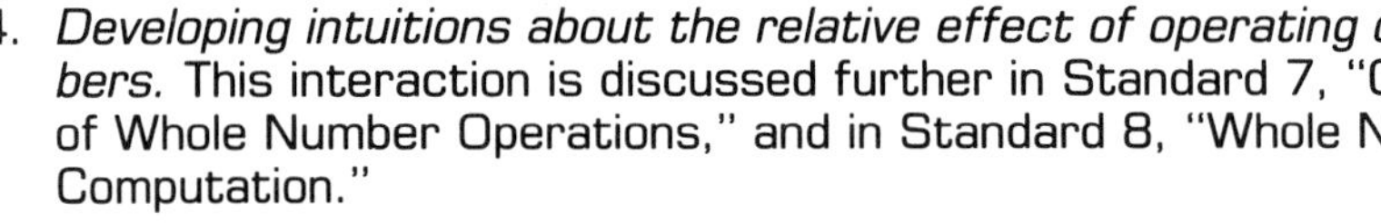

4. *Developing intuitions about the relative effect of operating on numbers.* This interaction is discussed further in Standard 7, "Concepts of Whole Number Operations," and in Standard 8, "Whole Number Computation."

5. *Developing referents for measures of common objects and situations in their environment.* For example, it is unrealistic for a fourth-grade child to be 316 cm tall or to weigh 8 kg, a loaf of bread doesn't cost $117, and the teacher is not ninety-six years old. A knowledge of reasonable ranges for such measures provides a basis for judging reasonableness of results.

Fig. 6.5

The following classroom example (fig. 6.5) focuses on the first two components of number sense, the meaning of 5 and relating 5 to its component parts.

"Make some designs using five toothpicks in each. Use numbers to tell me how your design is built from the toothpicks."

The "Guess My Number" example below helps children develop number-sense ideas regarding the relative magnitudes of larger numbers.

The teacher tapes five metersticks, marked in centimeters, end to end along the front of the room. The left endpoint is labeled 0 and the right endpoint 500. One student (the selector) silently selects a number between 0 and 500, and the others try to guess it.

If the first child guesses 400 and the selector says "Too high," then that child points to the 400 location. If the next child guesses 220 and the selector says "Too low," then that child points to the 220 mark. Guesses continue until the secret number is guessed. The two children pointing initially at 220 and 400 move closer together with each guess, always bracketing the range of possibilities.

The activity in figure 6.6 focuses on relative magnitudes for even larger numbers.

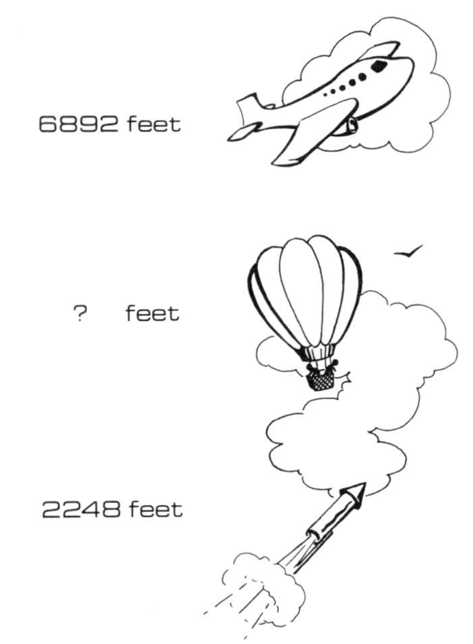

"About how high is the balloon?"

Fig. 6.6

STANDARD 7: CONCEPTS OF WHOLE NUMBER OPERATIONS

In grades K–4, the mathematics curriculum should include concepts of addition, subtraction, multiplication, and division of whole numbers so that students can—

- ***develop meaning for the operations by modeling and discussing a rich variety of problem situations;***
- ***relate the mathematical language and symbolism of operations to problem situations and informal language;***
- ***recognize that a wide variety of problem structures can be represented by a single operation;***
- ***develop operation sense.***

Focus

Understanding the fundamental operations of addition, subtraction, multiplication, and division is central to knowing mathematics. One essential component of what it means to understand an operation is recognizing conditions in real-world situations that indicate that the operation would be useful in those situations. Other components include building an awareness of models and the properties of an operation, seeing relationships among operations, and acquiring insight into the effects of an operation on a pair of numbers. These four components are aspects of *operation sense*. Children with good operation sense are able to apply operations meaningfully and with flexibility. Operation sense interacts with number sense and enables students to make thoughtful decisions about the reasonableness of results. Furthermore, operation sense provides a framework for the conceptual development of mental and written computational procedures.

Instruction on the meaning of operations focuses on concepts and relationships rather than on computation, which is the focus of Standard 8. Children need extensive informal experience with problem situations and language prior to explicit instruction and symbolic work with the operations. Thus, informal experiences with all four operations should begin in kindergarten and continue through grade 4. Instruction should help children connect their intuitions and informal language to operations, including the mathematical language and symbols of each operation. Children should encounter the four basic operations in a wide variety of problem structures. For example, in addition to problems with joining and separating structures, teachers should provide problems involving comparing and equalizing. Time devoted to conceptual development provides meaning and context to subsequent work on computational skills.

Maria has 5 cars. Bill has 8 cars. How many more cars does Bill have?

Anton, Juanita, and Booker want to share 6 cookies equally. How many cookies does each one get?

Fig. 7.1

Discussion

When most children enter school, they can use objects and counting to solve many kinds of problems. Class discussion of a wide variety of problems prepares students for explicit instruction on operations (the models, language, and symbols associated with an operation). Examples might include those in figure 7.1.

♦ ♦ ♦ ♦ ♦ ♦ ♦ ♦

Children draw upon their insight and intuition to represent and discuss these problems. This work helps link operations to many types of situations. Even after an operation has been introduced, this emphasis on linking it to appropriate situations should continue. For example, children might draw pictures and then tell or write stories about the equation as $18 \div 6 = \square$. This kind of activity emphasizes connections between mathematics and the real world and encourages children to recognize and use a variety of situations and problem structures.

Word problems also should be used to help children increase their recognition of the relationship between a single operation and problems with different structures. Although children initially might solve the following problem by drawing a picture, they should also see that because it involves separating a whole into equal parts, it can be solved using division.

Twenty-eight children are going on a picnic. Four children can ride in each car. How many cars are needed?

Connecting problem structures to operations should be emphasized throughout grades K–4 for both one-step and appropriate two-step problems. For example, multiplication is most commonly linked to the process of combining equal groups. Children also need to see that it relates to array, "times as many," and "combination" (e.g., three blouses, four skirts, how many outfits?) situations. An example of combination situations is finding the number of outfits that can be made with two blouses and three skirts.

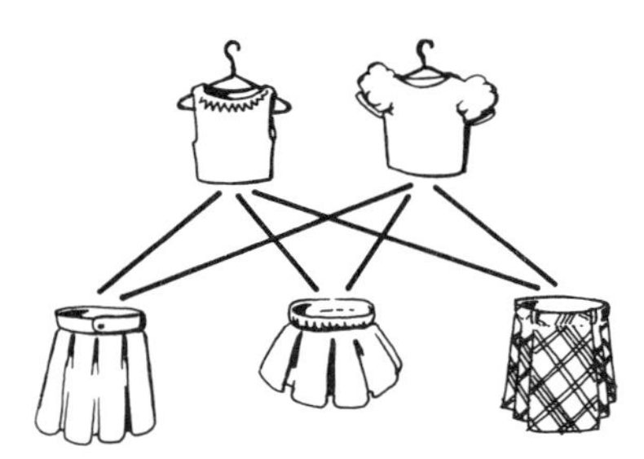

The language of basic operations, such as the terms *addend, sum, difference, factor, multiple, product,* and *quotient*, can be introduced and used informally in work with operations. The notions of factors and multiples can prompt interesting explorations. Children can find the factors of a number using tiles or graph paper. This can lead to an investigation of numbers that have only two factors (prime numbers) and numbers with two equal factors (square numbers).

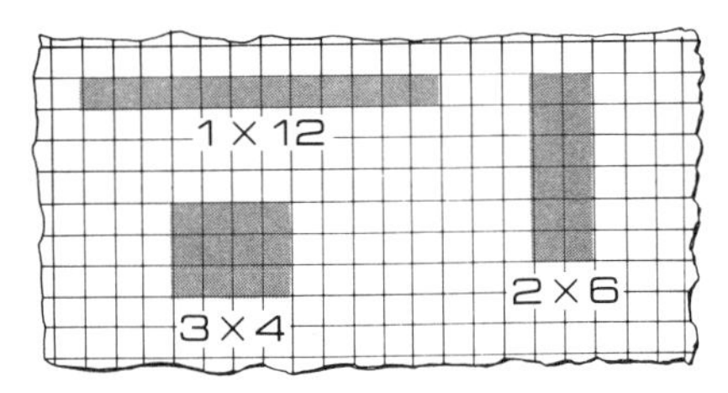

Factors of 12

Multiples of a number can be shaded on hundred charts. Children can then find numbers that are multiples of 2 and multiples of 3, and thus be introduced to the concept of common multiples. Calculators can be useful in exploring multiples of a number through repeated addition. After children become familiar with finding multiples of 3 (3, 6, 9, . . .), they can find how many threes make 30 using repeated addition and predict how many threes make 60. The calculator is used to check their predictions. Since work with concepts of operations does *not* emphasize the computing of answers, calculators are a valuable tool.

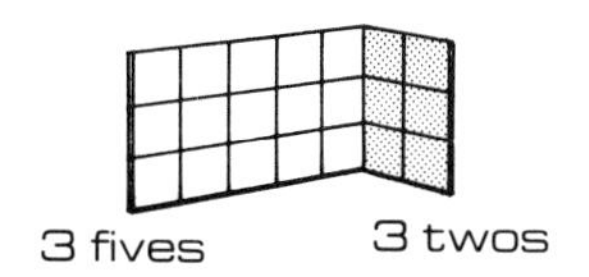

Fig. 7.2

Properties of an operation, a key component of operation sense, also can be explored. Children note that reversing the order of two addends does not change the sum, and they use this to solve $2 + 19$ by starting with 19 and counting two more. With graph paper, they can see that 3×7 can be found by adding 3 fives and 3 twos. See figure 7.2. Naming properties is not necessary.

Operation sense also involves relationships between operations. Addition and subtraction are related; for addition you find the whole, and for subtraction you find a part (see fig. 7.3).

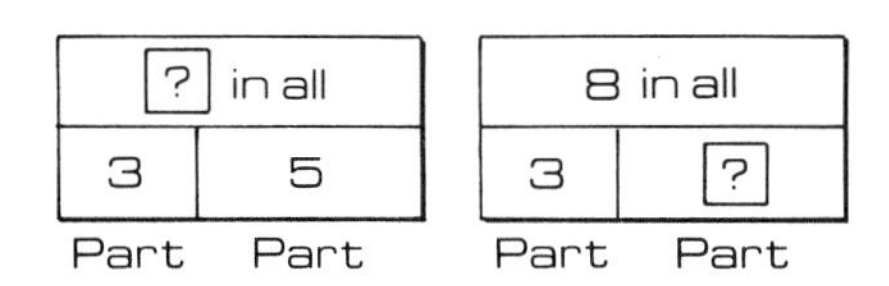

Fig. 7.3

Multiplication and division also have an inverse relationship. The relationships between addition and multiplication and between subtraction and division should be investigated.

Operation sense also involves acquiring insight and intuition about the effects of operations on two numbers. Adding 5 to 25, for example, produces a far smaller change in size than multiplying 25 by 5. Children should sense that the sum of two numbers, each of which is greater than 50, must be greater than 100. They can explore the effect of increasing one addend by 1 and decreasing the other addend by 1, and compare this to the corresponding results in multiplication:

$$\begin{array}{ccccc} 25 & + & 25 & = & 50 \\ \downarrow & & \downarrow & & \\ 26 & + & 24 & = & 50 \\ \downarrow & & \downarrow & & \\ 27 & + & 23 & = & 50 \end{array} \qquad \begin{array}{ccccc} 7 & \times & 7 & = & 49 \\ \downarrow & & \downarrow & & \\ 8 & \times & 6 & = & 48 \\ \downarrow & & \downarrow & & \\ 9 & \times & 5 & = & 45 \end{array}$$

Understandings of the relationships between operations can be used to extend work with equations. Children with a solid understanding of operations will be able to apply this knowledge to solve such equations as $15 + \square = 25$, $\square - 15 = 15$, and $\square \times 25 = 50$.

STANDARD 8: WHOLE NUMBER COMPUTATION

In grades K–4, the mathematics curriculum should develop whole number computation so that students can—

- ***model, explain, and develop reasonable proficiency with basic facts and algorithms;***
- ***use a variety of mental computation and estimation techniques;***
- ***use calculators in appropriate computational situations;***
- ***select and use computation techniques appropriate to specific problems and determine whether the results are reasonable.***

Focus

The purpose of computation is to solve problems. Thus, although computation is important in mathematics and in daily life, our technological age requires us to rethink how computation is done today. Almost all complex computation today is done by calculators and computers. In many daily situations, answers are computed mentally or estimates are sufficient, and paper-and-pencil algorithms are useful when the computation is reasonably straightforward. This standard addresses the importance of teaching children a variety of ways to compute, as well as the usefulness of calculators in solving problems containing large numbers or requiring complex computations. Related to this goal is the necessity of having reasonable expectations for proficiency with paper-and-pencil computation. Clearly, paper-and-pencil computation cannot continue to dominate the curriculum or there will be insufficient time for children to learn other, more important mathematics they need now and in the future.

By emphasizing underlying concepts, using physical materials to model procedures, linking the manipulation of materials to the steps of the procedures, and developing thinking patterns, teachers can help children master basic facts and algorithms and understand their usefulness and relevance to daily situations. This approach also promotes efficient learning of computational techniques and furthers the development of children's reasoning, mathematical insight, and confidence in their ability to do mathematics. Instruction also should emphasize a variety of ways to compute, the importance of checking whether computed results are reasonable, and the need to make appropriate decisions about how to compute in a problem situation. An awareness that computation is learned and used to attain some goal develops when problem situations and computations are explicitly linked throughout all aspects of work with computations.

Discussion

Strong evidence suggests that conceptual approaches to computation instruction result in good achievement, good retention, and a reduction in the amount of time children need to master computational skills. Furthermore, many of the errors children typically make are less prevalent.

Helping children develop thinking strategies for learning basic facts enables them to understand relationships and to reason mathematically. Figure 8.1 shows two examples.

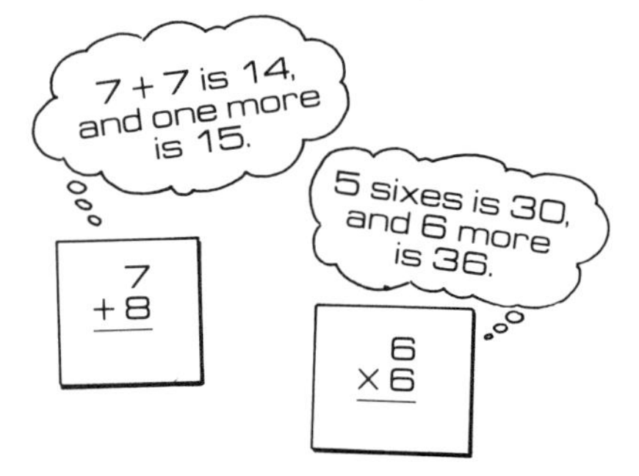

Fig. 8.1

A developmental approach to computation fosters a problem-solving atmosphere in which children are actively involved in using materials, discussing their work, validating solutions, and raising questions. Placing computation in a problem-solving context motivates students to learn computational skills and serves as an impetus for the mastery of paper-and-pencil algorithms. The initial use of physical materials, such as base-ten blocks or bundling sticks, can be carefully connected to concrete models and, finally, to symbolic work. Figure 8.2 illustrates the connections that can be made between concrete materials and a paper-and-pencil algorithm.

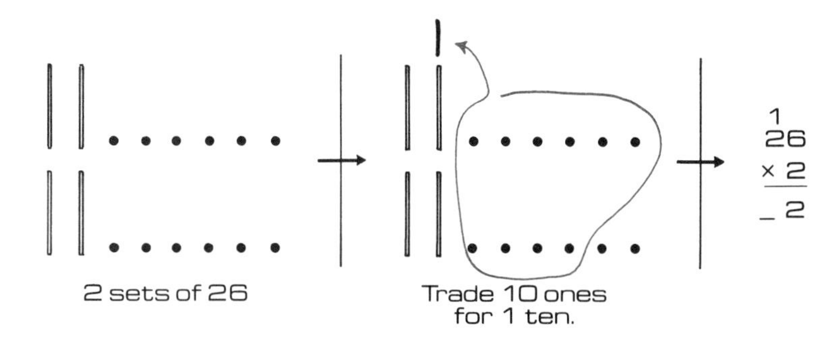

Fig. 8.2

Mental computation and estimation offer exciting opportunities for making computation more dynamic and for developing insights into number relationships. Figure 8.3 illustrates several thinking patterns.

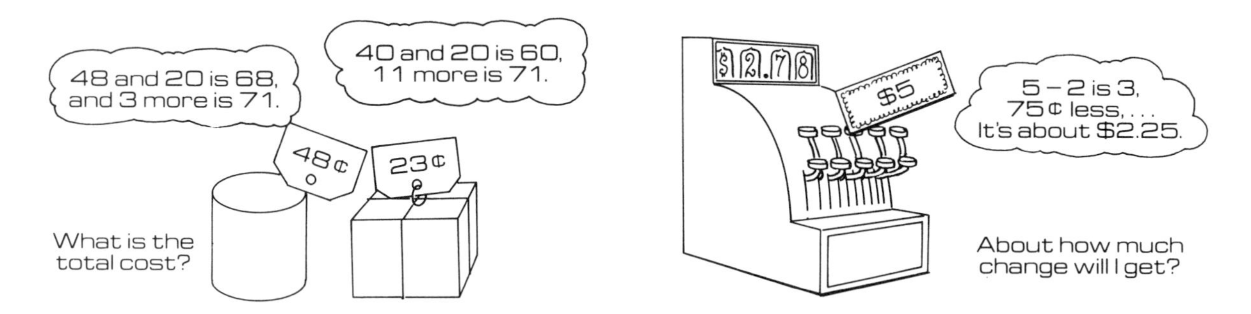

Fig. 8.3

Children need more time to explore and to invent alternative strategies for computing mentally. Both mental computation and estimation should be ongoing emphases that are integrated throughout all computational work. Estimation is discussed further in Standard 5.

The frequent use of calculators, mental computation, and estimation helps children develop a more realistic view of computation and enables them to be more flexible in their selection of computing methods. Calculators should be used to solve problems that require tedious calculations. Estimation and reasonableness of results need particular emphasis when students are using calculators. The following example illustrates how to design various problems so that students must check the reasonableness of their results once they have completed their work with a calculator.

Three fourth-grade teachers at Park City Elementary School decided to take all their students on a picnic. Mr. Clark spent $26.94 for refreshments. Since the three teachers wanted to share the cost of the picnic, Mr. Clark used his calculator to determine that each teacher should pay him $13.47. Is his answer reasonable? Explain.

After estimating, the students concluded that Mr. Clark was wrong because 27 divided by 3 is 9; thus, $9.00 is about what he should collect from each teacher, not $13.47.

In this example, estimation showed that the teacher's answer was in error.

Calculators also can be used as an effective instructional tool for teaching computational skills. For example:

Target Addition is a calculator game for reinforcing the recall of basic facts and mental arithmetic. After clearing the calculator's memory, two children select a target, such as 23, and take turns entering a number from 1 to 5. Each new sum is put into the memory by pressing the M+ key. A player who thinks the target number is in the memory just after his or her turn presses the memory-recall key to check.

Children should also be given many opportunities to decide whether they need an exact answer and how they will complete a computation. See figure 8.4.

How much in all?

Fig. 8.4

Is $5 enough to buy these four things?

Rethinking the role of computation. The approach to computation taken in this standard requires educators to rethink traditional scope-and-sequence decisions. If they are to meet the comprehensive curricular goals articulated in the K–4 standards, for example, teachers must reduce the time and the emphasis they devote to computation and focus instead on the other mathematical topics and perspectives that are proposed.

Besides paper-and-pencil computation, children should learn when and how to use calculators and various mental arithmetic and estimation procedures. Calculators enable children to compute to solve problems beyond their paper-and-pencil skills. Mental computation and estimation techniques can be developed prior to, and in connection with, paper-and-pencil skills. It is inconsistent with the *Standards* to isolate paper-and-pencil procedures by focusing on them for an extended time prior to the introduction of other computing methods; this traditional practice suggests to children that computing means using paper-and-pencil methods.

Reasonable expectations for computation. Premature expectations for students' mastery of computational procedures not only cause poor initial learning and poor retention but also require that large amounts of instructional time be spent on teaching and reteaching basic skills. More important, the instructional focus centers on memorizing facts and rules for carrying out procedures rather than on the thoughtful use of operations and number relationships.

Children should master the basic facts of arithmetic that are essential components of fluency with paper-and-pencil and mental computation and with estimation. At the same time, however, mastery should not be expected too soon. Children will need many exploratory experiences and the time to identify relationships among numbers and efficient thinking strategies to derive the answers to unknown facts from known facts. Practice designed to improve speed and accuracy should be used, but only under the right conditions; that is, practice with a cluster of facts should be used only after children have developed an efficient way to derive the answers to those facts.

It is important for children to learn the sequence of steps—and the reasons for them—in the paper-and-pencil algorithms used widely in our culture. Thus, instruction should emphasize the meaningful development of these procedures, not speed of processing. The teaching of addition, subtraction, and multiplication algorithms should integrate renaming and no-renaming situations, and problems with remainders should be integrated throughout division. This approach is more efficient and eliminates some misconceptions that often occur.

Exploratory experiences in preparation for paper-and-pencil computation give children the opportunity to develop underlying concepts related to partitioning numbers, operating on the parts, and combining the results. Many such experiences can be provided in the context of using place-value materials, computing mentally, or performing computational estimation. Only after these ideas are carefully linked to paper-and-pencil procedures is it appropriate to devote time to developing proficiency by providing practice. Although the exploration of computation with larger numbers is appropriate, excessive amounts of time should not be devoted to proficiency.

Success is possible for almost all children when they receive careful instruction. Still, teachers should be sensitive to problems individual children might have and should be prepared to use a variety of methods to teach and assess computational knowledge.

STANDARD 9: GEOMETRY AND SPATIAL SENSE

In grades K–4, the mathematics curriculum should include two- and three-dimensional geometry so that students can—

- ***describe, model, draw, and classify shapes;***
- ***investigate and predict the results of combining, subdividing, and changing shapes;***
- ***develop spatial sense;***
- ***relate geometric ideas to number and measurement ideas;***
- ***recognize and appreciate geometry in their world.***

Focus

Geometry is an important component of the K–4 mathematics curriculum because geometric knowledge, relationships, and insights are useful in everyday situations and are connected to other mathematical topics and school subjects. Geometry helps us represent and describe in an orderly manner the world in which we live. Children are naturally interested in geometry and find it intriguing and motivating; their spatial capabilities frequently exceed their numerical skills, and tapping these strengths can foster an interest in mathematics and improve number understandings and skills.

Spatial understandings are necessary for interpreting, understanding, and appreciating our inherently geometric world. Insights and intuitions about two- and three-dimensional shapes and their characteristics, the interrelationships of shapes, and the effects of changes to shapes are important aspects of spatial sense. Children who develop a strong sense of spatial relationships and who master the concepts and language of geometry are better prepared to learn number and measurement ideas, as well as other advanced mathematical topics.

In learning geometry, children need to investigate, experiment, and explore with everyday objects and other physical materials. Exercises that ask children to visualize, draw, and compare shapes in various positions will help develop their spatial sense. Although a facility with the language of geometry is important, it should not be the focus of the geometry program but rather should grow naturally from exploration and experience. Explorations can range from simple activities to challenging problem-solving situations that develop useful mathematical thinking skills.

Evidence suggests that the development of geometric ideas progresses through a hierarachy of levels. Students first learn to recognize whole shapes and then to analyze the relevant properties of a shape. Later they can see relationships between shapes and make simple deductions. Curriculum development and instruction must consider this hierarchy because although learning can occur at several levels simultaneously, the learning of more complex concepts and strategies requires a firm foundation of basic skills.

Discussion

Geometry gives children a different view of mathematics. As they explore patterns and relationships with models, blocks, geoboards, and

♦ ♦ ♦ ♦ ♦ ♦ ♦ ♦

graph paper, they learn about the properties of shapes and sharpen their intuitions and awareness of spatial concepts. Children's geometric ideas can be developed by having them sort and classify models of plane and solid figures, construct models from straws, make drawings, and create and manipulate shapes on a computer screen. Folding paper cutouts or using mirrors to investigate lines of symmetry are other ways for children to observe figures in a variety of positions, become aware of their important properties, and compare and contrast them. Related experiences help children avoid simplistic and misleading ideas about shapes, such as that implied by one child's observation, "This is an upside-down triangle."

Children can be taught to internalize characteristics of shapes and then translate those internal ideas into descriptions and models by feeling an object inside a paper bag, identifying the object, creating the shape on a geoboard, and naming the object's shape.

Children can also follow verbal directions, such as "Draw [or put on a geoboard] a shape that has four sides and two right angles." Follow-up discussions help children understand the conditions necessary to define a shape. These experiences allow children to develop more complete understandings about shapes and their properties and to build the vocabulary of geometry in a natural manner.

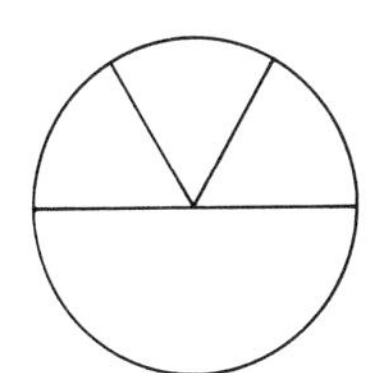

Fig. 9.1

Spatial sense is an intuitive feel for one's surroundings and the objects in them. To develop spatial sense, children must have many experiences that focus on geometric relationships; the direction, orientation, and perspectives of objects in space; the relative shapes and sizes of figures and objects; and how a change in shape relates to a change in size. These experiences depend on a child's ability to follow directions that use words like *above, below,* and *behind* and to progress to such activities as using a computer to reproduce a pattern-block design. When children examine the result of combining two shapes to form a new shape, predict the effect of changing the number of sides of a shape, draw a shape after it has been rotated a quarter or half turn, or explore what happens when the dimensions of a shape are changed, they acquire a deeper understanding of shapes and their properties. Such activities promote spatial sense.

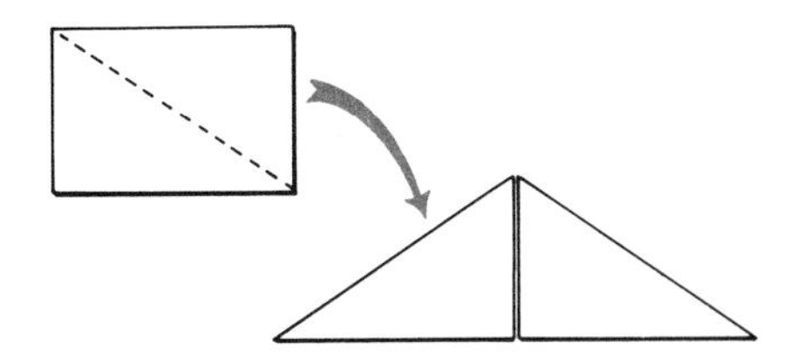

Fig. 9.2

Drawing and sketching shapes is an important part of developing spatial sense. The following spatial-visualization tasks illustrate one productive activity. A figure is displayed on an overhead projector for two to three seconds (fig. 9.1) and then children try to draw the figure. The original figure is again briefly displayed, and children make a second attempt at drawing it. Discussion about what the children saw is also important.

Children can cut paper shapes and make new shapes from the parts (fig. 9.2).

When children hold a long loop of yarn so that each hand serves as a vertex, they can explore the effect of changing the size of an angle, or increasing the number of sides while the perimeter is unchanged. See figure 9.3.

Fig. 9.3

Another activity that promotes spatial sense is to have children decide which two-dimensional patterns can be folded to produce a three-dimensional shape. See figure 9.4.

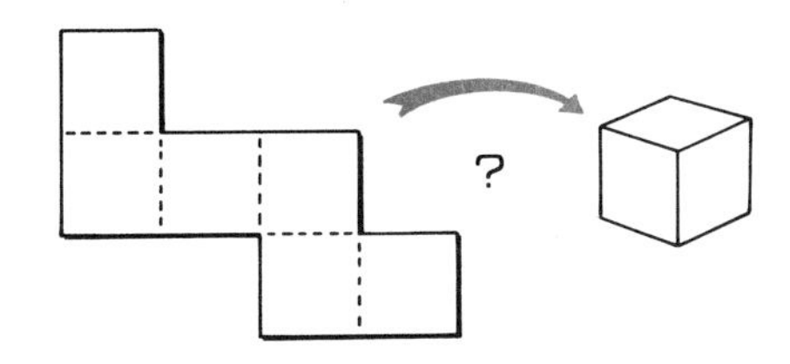

Fig. 9.4

Geometry contributes to the development of number and measurement concepts. Two-dimensional regions are subdivided into congruent parts to teach fraction and decimal concepts. Geometric ideas and a number line are useful models for teaching rounding. For example, 438 is *between* 400 and 500; it is *closer to* 400 because it is less than *halfway*.

Many geometric skills and concepts are essential to the process of problem solving. For example, a primary problem-solving strategy is drawing a picture or diagram, which is, in many situations, a geometric representation of the problem. Three sample problems illustrate this point. See figures 9.5 and 9.6.

In summary, children should have many opportunities to explore geometry in two and three dimensions, to develop their sense of space and relationships in space, and to solve problems that involve geometry and its application to other topics in mathematics or to other fields.

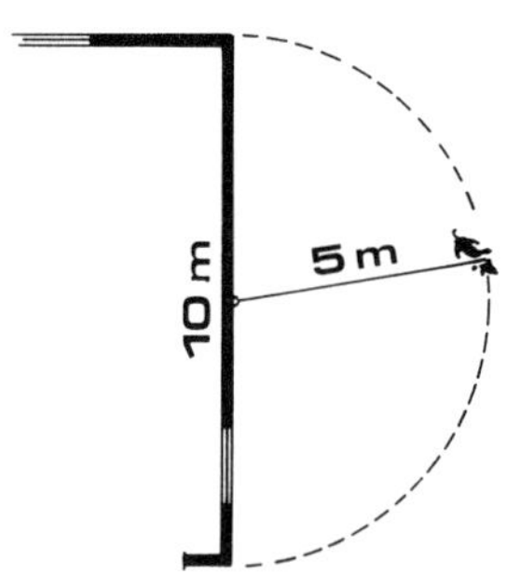

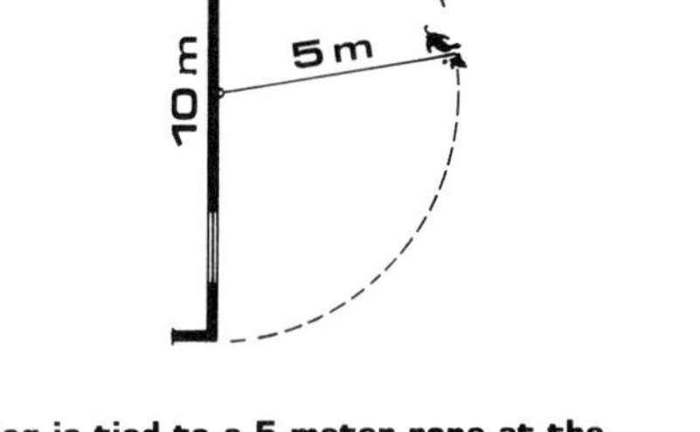

A dog is tied to a 5-meter rope at the middle of the side of a garage. The side of the garage is 10 meters long. Make a sketch and use centimeter grid paper to estimate the area and shape of the ground on which the dog can walk.

Fig. 9.5

This is part of a closed figure. Complete it so that it has two lines of symmetry.

Make another shape that intersects this one to form a rectangle.

Fig. 9.6

STANDARD 10: MEASUREMENT

In grades K–4, the mathematics curriculum should include measurement so that students can—

- ***understand the attributes of length, capacity, weight, mass, area, volume, time, temperature, and angle;***
- ***develop the process of measuring and concepts related to units of measurement;***
- ***make and use estimates of measurement;***
- ***make and use measurements in problem and everyday situations.***

Focus

Measurement is of central importance to the curriculum because of its power to help children see that mathematics is useful in everyday life and to help them develop many mathematical concepts and skills. Measuring is a natural context in which to introduce the need for learning about fractions and decimals, and it encourages children to be actively involved in solving and discussing problems.

Instruction at the K–4 level emphasizes the importance of establishing a firm foundation in the basic underlying concepts and skills of measurement. Children need to understand the attribute to be measured as well as what it means to measure. Before they are capable of such understanding, they must first experience a variety of activities that focus on comparing objects directly, covering them with various units, and counting the units. Premature use of instruments or formulas leaves children without the understanding necessary for solving measurement problems.

Estimation should be emphasized because it helps children understand the attributes and the process of measuring as well as gain an awareness of the sizes of units. Everyday situations in which only an estimate is required should be included. Since measurements are not exact, children should realize that it is often appropriate, for example, to report a measurement as between eight and nine centimeters or about three hours.

As measurement concepts and skills are introduced, they should be integrated throughout mathematics and other curriculum areas. Not only will this enhance other topics but it will also give children opportunities to develop and retain measurement concepts and skills.

Discussion

The approach advocated in this standard will give children a firm foundation that enables them to use any measurement system. The first step in building this foundation is understanding an object's many measurable attributes, such as those illustrated by a cereal box. See figure 10.1.

Children begin to develop an understanding of such attributes through experiences like those in figure 10.2, in which they make decisions about the sizes of objects by looking, feeling, or comparing objects directly. These experiences also provide the opportunity in a natural way to build much of the vocabulary associated with measurement.

How much does it hold? (capacity)
How tall is it? (length)
How large is the front? (area)
How heavy is it? (mass or weight)
How far around is the border? (length or perimeter)

Fig. 10.1

Lift these two rocks. Which is heavier? Use the balance to check.

Draw a line in the sand that is shorter than the stick.

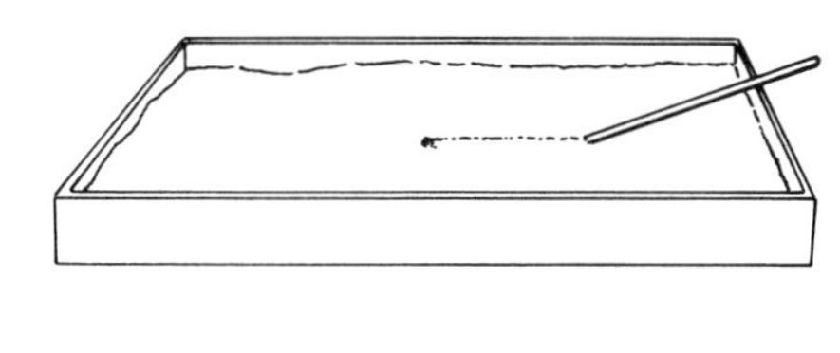

Compare the angles of these figures. Which is largest?

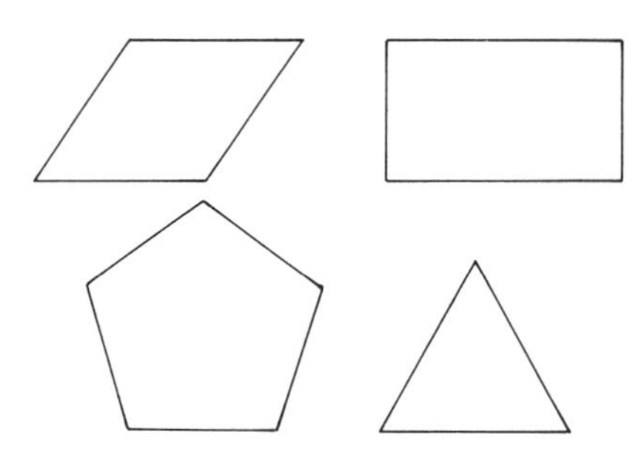

Fig. 10.2

The process of measuring is identical for any attribute: Choose a unit, compare that unit to the object, and report the number of units. The number of units can be determined by counting, by using an instrument, or by using a formula. In the examples in figure 10.3, the number of area units is determined by counting and the number of length units is determined with a ruler.

Measurement Process	***Attribute***	
	Area	***Length***
Choose a unit.	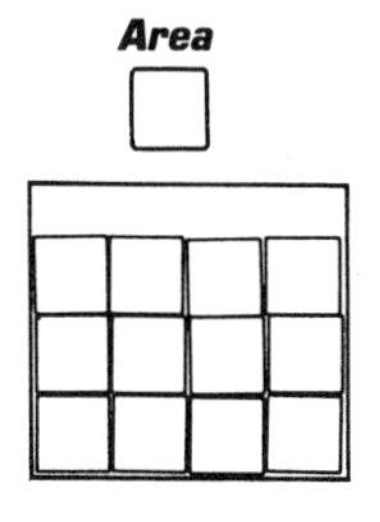	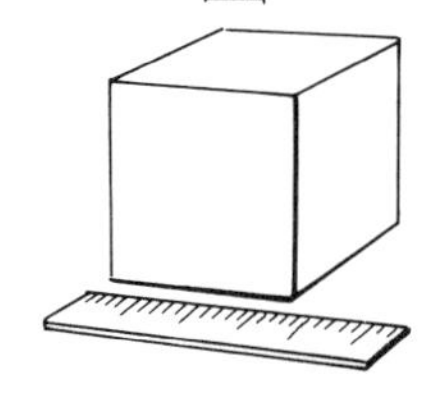
Compare the unit to the object.		
Report the number and the unit.	**"It's more than 12 units."**	**"It's between 19 cm and 20 cm."**

Fig. 10.3

Many important understandings are associated with a unit of measure. The choice of a unit is arbitrary, but it must have the same attribute as that which is being measured. That is, a unit of area must be selected to measure area, a unit of weight to measure weight, and so forth. The size of an appropriate unit depends on the size of the object or the desired precision of the measurement. For example, it is appropriate to choose a large container as a unit to measure the capacity of a bathtub and a small container to measure the capacity of a teacup. Children can also explore the relationship between the size of a unit and the number of units it takes to measure an object, as shown in figure 10.4.

The table is about six straws long. The table is about eight crayons long.

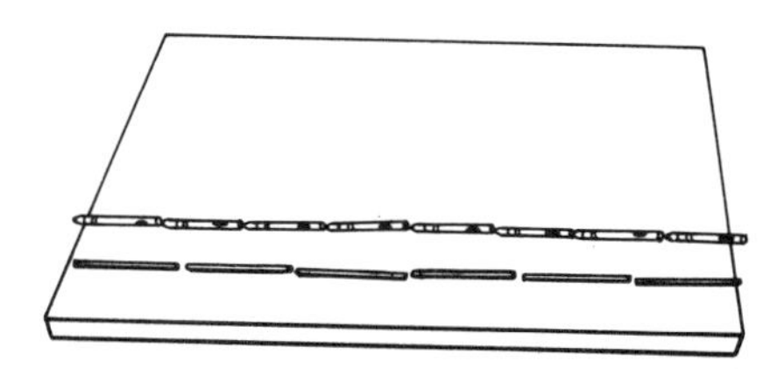

It takes fewer straws than crayons because the straws are longer.

Fig. 10.4

If children's initial explorations use nonstandard units, they will develop some understandings about units and come to recognize the necessity of standard units in order to communicate. Children can build an awareness of the approximate size of a standard unit through activities in which they find objects with a length of 1 meter, a mass of 1 gram, or a capacity of 1 liter. Measuring the same object with different standard units provides the background for learning the basic relationships between units and conversions at the middle grades. (For example, children can report the height of a door as 2 meters and 10 centimeters, or as 210 centimeters.) Such work also helps children become aware of the approximate nature of measurement.

Estimation activities should be integrated throughout measurement, including those that ask for an estimate of the measure of an object

♦ ♦ ♦ ♦ ♦ ♦ ♦ ♦

(About how large is the angle?) and those that ask for an object of a given measure (Find a piece of paper that is five centimeters long). The computer should not be overlooked as a tool that encourages estimation. When drawing figures on a computer, one often finds it necessary to estimate the length of a line or the result of a turn of a given number of degrees. Activities also should be provided that encourage the use of such estimation strategies as chunking (estimating the whole by estimating its parts).

Children can see the usefulness of measurement if classroom experiences focus on measuring real objects, making objects of given sizes, and estimating measurements. Textbook experiences cannot substitute for activities that use measurement to answer questions about real problems.

STANDARD 11: STATISTICS AND PROBABILITY

In grades K–4, the mathematics curriculum should include experiences with data analysis and probability so that students can—

- ***collect, organize, and describe data;***
- ***construct, read, and interpret displays of data;***
- ***formulate and solve problems that involve collecting and analyzing data;***
- ***explore concepts of chance.***

Focus

Collecting, organizing, describing, displaying, and interpreting data, as well as making decisions and predictions on the basis of that information, are skills that are increasingly important in a society based on technology and communication. These processes are particularly appropriate for young children because they can be used to solve problems that often are inherently interesting, represent significant applications of mathematics to practical questions, and offer rich opportunities for mathematical inquiry. The study of statistics and probability highlights the importance of questioning, conjecturing, and searching for relationships when formulating and solving real-world problems.

A spirit of investigation and exploration should permeate statistics instruction. Children's questions about the physical world can often be answered by collecting and analyzing data. After generating questions, they decide what information is appropriate and how it can be collected, displayed, and interpreted to answer their questions. The analysis and evaluation that occur as children attempt to draw conclusions about the original problem often lead to new conjectures and productive investigations. This entire process broadens children's views of mathematics and its usefulness.

Statistics and probability are important links to other content areas, such as social studies and science. They also can reinforce communication skills as children discuss and write about their activities and their conclusions. Within mathematics, these topics regularly involve the uses of number, measurement, estimation, and problem solving.

Discussion

This standard recognizes the importance of having all students develop an awareness of the concepts and processes of statistics and probability. The curriculum must emphasize that statistics is more than reading and interpreting graphs: It is describing and interpreting the world around us with numbers, and it is a tool for solving problems. Children need to recognize that many kinds of data come in many forms and that collecting, organizing, displaying, and thinking about them can be done in many ways.

In the early grades, actual objects should be displayed so that their characteristics can be observed and discussed. In this work, each unit used on scales for graphs should represent 1. Later, pictorial and symbolic graphs should be constructed and discussed; by grades 3 and 4, children should be able to use scales representing other units, such as 2, 5, or 10. Computer graphing programs make a wide variety of explorations accessible to older children after they have learned to create graphs on their own. Exercises like the following encourage children to use pictures and symbols to characterize and group objects.

Pretend we own a children's shoe store. We need to know whether to have more cloth or more leather shoes for sale in our store. What could we do to decide?

The children might decide to make a floor graph with one shoe from each child as a way of determining the number of cloth and leather shoes in their class. Questions to guide students' activities can include these: Are there more cloth shoes or leather shoes? Are the two numbers close? Should we have about the same number of cloth and leather shoes in our store?

Children should learn that data can be displayed in different ways and that depending on the question being asked, one type of display might be more appropriate than another. A variety of early experiences helps children build a foundation for creating conventional graphs. See figure 11.1.

Which display would we use to find out the kind of ice cream that Molly likes? Which display would we use to find out which flavor is the most popular?

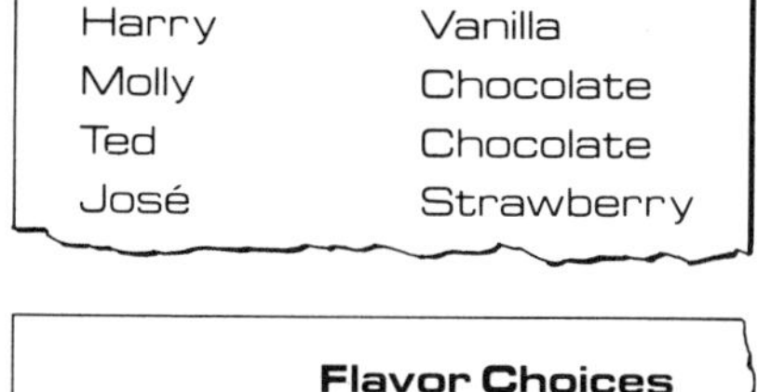

Survey	
Harry	Vanilla
Molly	Chocolate
Ted	Chocolate
José	Strawberry

Flavor Choices					
Vanilla	X	X	X		
Chocolate	X	X	X	X	X
Strawberry	X	X			

What is the favorite flavor of ice cream in our class?

Fig. 11.1

A class or group project conducted over time enables the students to make predictions and modify them as more data are collected.

Suppose, for example, that children are interested in comparing the temperatures in their hometown with the temperatures in two other cities. They can obtain pertinent data from such sources as newspapers or television. They can participate in making decisions about what questions to ask; what data to collect; and how to collect, organize, and display them for others to see and interpret. See figure 11.2.

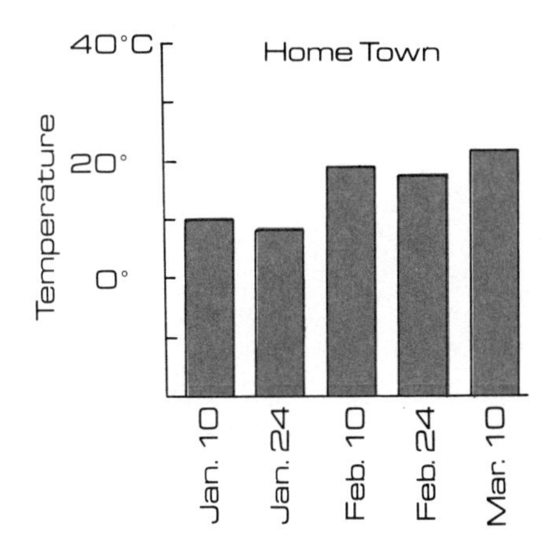

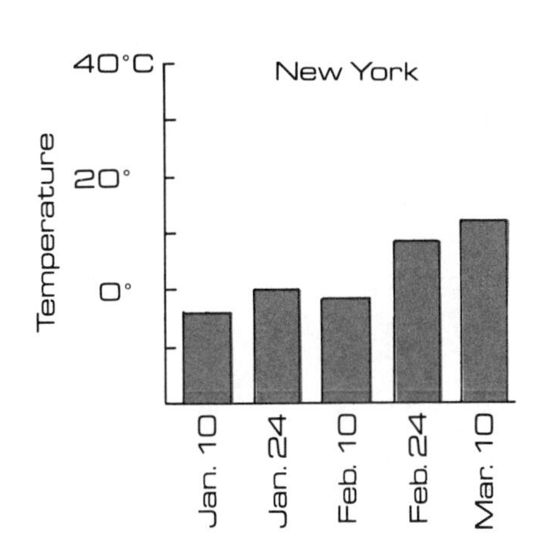

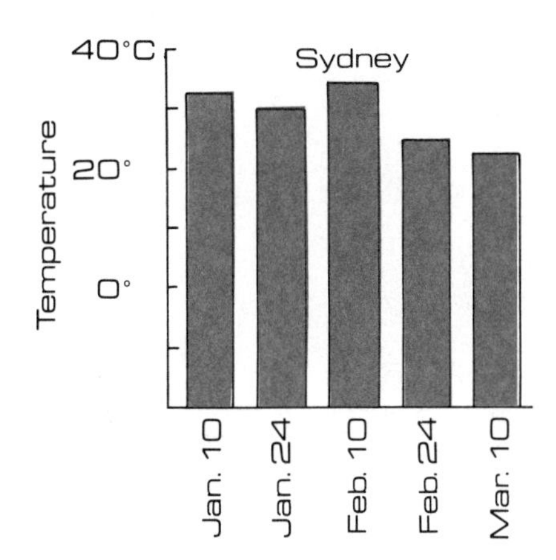

Fig. 11.2

Again, teachers might find that questions like the following are helpful in guiding students' efforts.

What patterns do you notice? Did anything about the data surprise you? What temperature do you predict for each city on 24 March? Will these temperature trends continue through 24 December? Why do you think New York is getting warmer and Sydney is getting colder?

Children of this age will also enjoy and profit from the exploration of chance. This pursuit should have the same investigative flavor as that recommended for statistics, as illustrated by the following activity involving spinners (fig. 11.3).

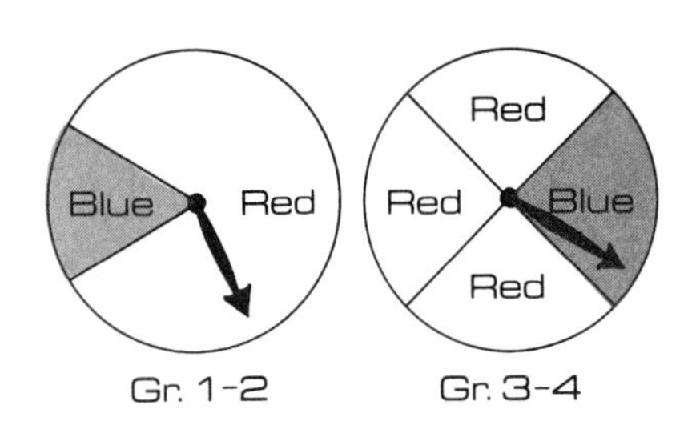

Fig. 11.3

Is red or blue more likely? How likely is yellow? How likely is getting either red or blue? If we spin twelve times, how many blues might we expect to get?

The following game combines an exploration of probability with data analysis:

Cut a hole slightly smaller than the size of a craft bead in the lid of an opaque bottle. Secretly place ten beads of two colors in the bottle (e.g., three red and seven blue). Two teams of children alternate turns; each team uses a number line that goes from 0 to 10. The starting point is 5. On each turn, the team decides whether red or blue will indicate movement toward 10. The members of each team take turns shaking the bottle upside down until one bead falls out. If the bead is of the color predicted, the team moves forward one space. If not, it moves back one space. The first team to reach either 0 or 10 wins. Children should be encouraged to keep a record of the colors that appear for later analysis and discussion.

At the conclusion of the game, each team guesses the number of beads of each color. The first team to identify correctly the number of each color gets to change the distribution of the two colors in the bottle for the next round.

This game allows students to explore many aspects of probability and gather and analyze data in a problem-solving atmosphere. Discussions following the game can include the concepts of events that are likely, events that are certain, and common perceptions of "luck."

STANDARD 12: FRACTIONS AND DECIMALS

In grades K–4, the mathematics curriculum should include fractions and decimals so that students can—

- ***develop concepts of fractions, mixed numbers, and decimals;***
- ***develop number sense for fractions and decimals;***
- ***use models to relate fractions to decimals and to find equivalent fractions;***
- ***use models to explore operations on fractions and decimals;***
- ***apply fractions and decimals to problem situations.***

Focus

Fractions and decimals represent a significant extension of children's knowledge about numbers. When children possess a sound understanding of fraction and decimal concepts, they can use this knowledge to describe real-world phenomena and apply it to problems involving measurement, probability, and statistics. An understanding of fractions and decimals broadens students' awareness of the usefulness and power of numbers and extends their knowledge of the number system. It is critical in grades K–4 to develop concepts and relationships that will serve as a foundation for more advanced concepts and skills.

The K–4 instruction should help students understand fractions and decimals, explore their relationship, and build initial concepts about order and equivalence. Because evidence suggests that children construct these ideas slowly, it is crucial that teachers use physical materials, diagrams, and real-world situations in conjunction with ongoing efforts to relate their learning experiences to oral language and symbols. This K–4 emphasis on basic ideas will reduce the amount of time currently spent in the upper grades in correcting students' misconceptions and procedural difficulties.

Discussion

All work at the K–4 level should involve fractions that are useful in everyday life, that is, fractions that can be easily modeled. Initial work with fractions should draw on children's experiences in sharing, such as asking four children to share a candy bar. The concept of a unit and its subdivision into equal parts is fundamental to understanding fractions and decimals, whether the quantity to be divided is a rectangular candy bar, a handful of jelly beans, or a piece of licorice. Initial instruction needs to emphasize oral language (one-fourth, two-thirds) and connect it to the models. Many productive activities can be used for initial instruction, such as folding paper strips into equal parts and describing the kind of parts (e.g., fifths) and the amount being considered (e.g., two-fifths).

In another activity, students construct a whole when given a part (fig. 12.1).

If this piece is one-fourth, make a whole.

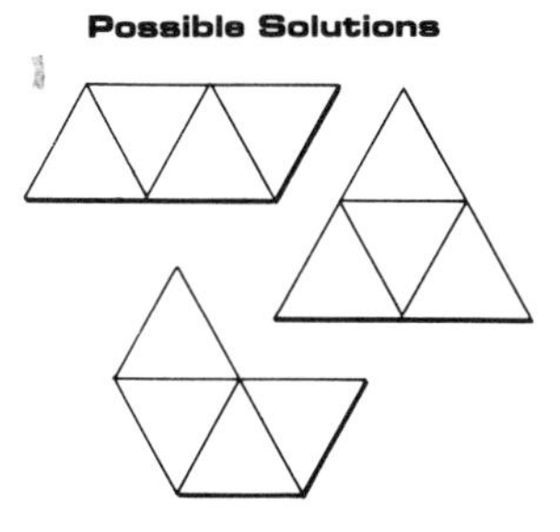

Fig. 12.1

Counting forward and backward by unit fractions ($1/2$, $1/3$, $1/4$, etc.) helps children build a strong awareness of fraction sequences and prepares them for both mental and paper-and-pencil computation. One relevant, thought-provoking activity appears in figure 12.2.

Divide the class into two groups. Let one group be the "mixed" group and the other the "improper" group. Have each group count the number of thirds shown:

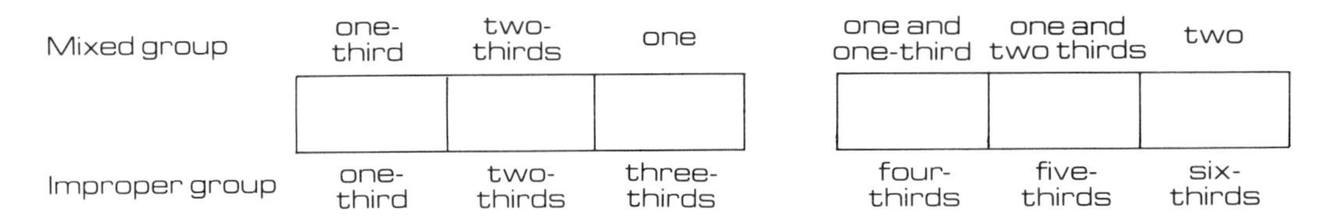

Fig. 12.2

Fraction symbols, such as $\frac{1}{4}$ and $\frac{3}{2}$, should be introduced only after children have developed the concepts and oral language necessary for symbols to be meaningful and should be carefully connected to both the models and oral language.

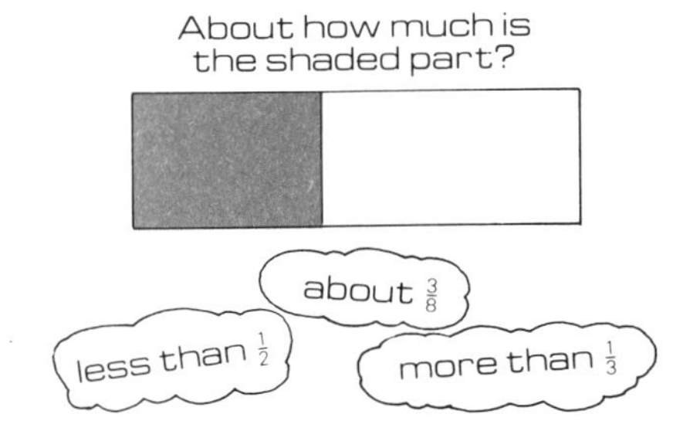

Fig. 12.3

An awareness of the relative size of fractions fosters number sense and enhances basic understandings. The following activity (see fig. 12.3) helps children think about the quantity represented by a fraction.

Children need to use physical materials to explore equivalent fractions and compare fractions. For example, with folded paper strips, children can easily see that $1/2$ is the same amount as $3/6$ and that $2/3$ is smaller than $3/4$.

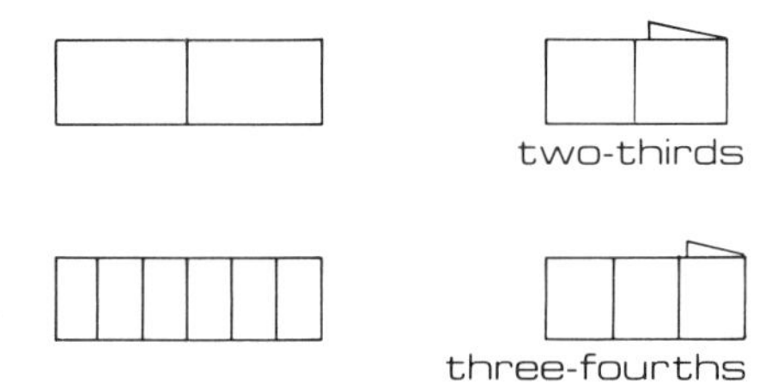

Children also should use reasoning to determine that $1/5$ is larger than $1/8$ or $1/10$ since fifths are larger than eights or tenths. Students should recognize that, for example, $3/4$ is between $1/2$ and 1 and that $1/3$ is large compared to $1/10$, about the same size as $1/4$, and small compared to $5/6$. They can also explore fractions that are close to 0, close to $1/2$, or close to 1, as in figure 12.4. Experiences with the relative size of numbers promote the development of number sense.

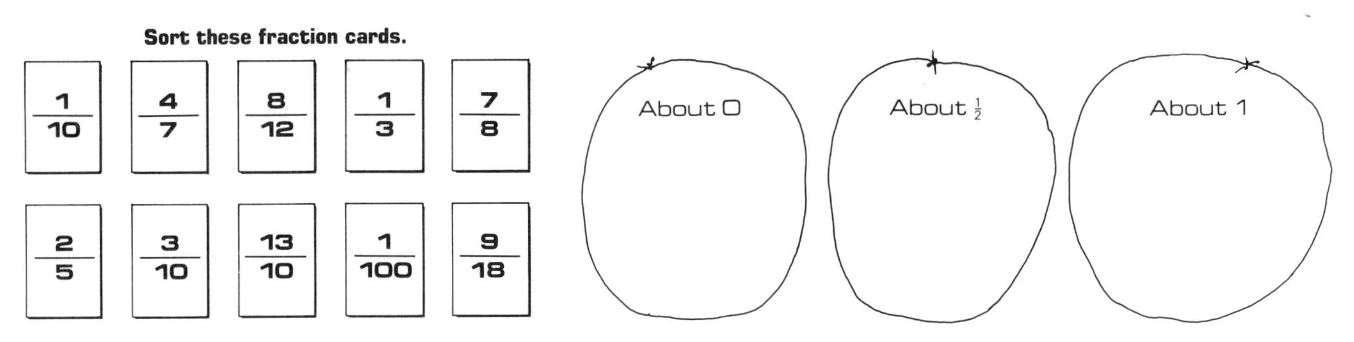

Fig. 12.4

♦ ♦ ♦ ♦ ♦ ♦ ♦ ♦

Physical materials should be used for exploratory work in adding and subtracting basic fractions, solving simple real-world problems, and partitioning sets of objects to find fractional parts of sets and relating this activity to division. For example, children learn that ⅓ of 30 is equivalent to "30 divided by 3," which helps them relate operations with fractions to earlier operations with whole numbers.

In grades K–4, children begin to encounter decimals in many situations—with calculators and metric measures, in tables of data, and in such daily activities as using a digital stopwatch. Thus, the curriculum needs to emphasize the development of decimal concepts.

The approach to decimals should be similar to work with fractions, namely, placing a strong and continued emphasis on models and oral language and then connecting this work with symbols. This is necessary if students are to make sense of decimals and use them insightfully. Exploring ideas of tenths and hundredths with models can include preliminary work with equivalent decimals (fig. 12.5), counting sequences, the comparing and ordering of decimals, and addition and subtraction.

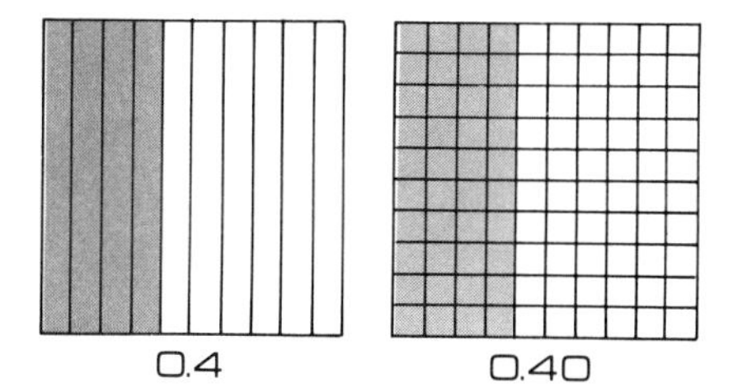

Fig. 12.5

Decimal instruction should include informal experiences that relate fractions to decimals so that students begin to establish connections between the two systems. For example, if students recognize that ½ is the same amount as 0.5, they can use this relationship to determine that 0.4 and 0.45 are a little less than ½ and that 0.6 and 0.57 are a little more than ½. Such activities help children develop number sense for decimals.

STANDARD 13: PATTERNS AND RELATIONSHIPS

In grades K–4, the mathematics curriculum should include the study of patterns and relationships so that students can—

- ***recognize, describe, extend, and create a wide variety of patterns;***
- ***represent and describe mathematical relationships;***
- ***explore the use of variables and open sentences to express relationships.***

Focus

Patterns are everywhere. Children who are encouraged to look for patterns and to express them mathematically begin to understand how mathematics applies to the world in which they live. Identifying and working with a wide variety of patterns help children to develop the ability to classify and organize information. Relating patterns in numbers, geometry, and measurement helps them understand connections among mathematical topics. Such connections foster the kind of mathematical thinking that serves as a foundation for the more abstract ideas studied in later grades.

From the earliest grades, the curriculum should give students opportunities to focus on regularities in events, shapes, designs, and sets of numbers. Children should begin to see that regularity is the essence of mathematics. The idea of a functional relationship can be intuitively developed through observations of regularity and work with generalizable patterns.

Physical materials and pictorial displays should be used to help children recognize and create patterns and relationships. Observing varied representations of the same pattern helps children identify its properties. The use of letters and other symbols in generalizing descriptions of these properties prepares children to use variables in the future. This experience builds readiness for a generalized view of mathematics and the later study of algebra.

Discussion

A classroom can become a rich environment in which to study patterns. A rug, an afghan, or quilt; various wallpaper borders used to identify boxes of materials; designs painted on the window panes; carefully selected pictures; and even the arrangement of furniture are examples of regularity and patterns that children can recognize and describe (see fig. 13.1).

Fig. 13.1

Regularities can be as simple as an explicit recognition that each child has two eyes. As each child in turn stands, the number of eyes of those standing can be recorded (2, 4, 6, . . .) and then represented with tiles to emphasize that even numbers come in pairs, whereas each odd number is an even number and one more. See figure 13.2.

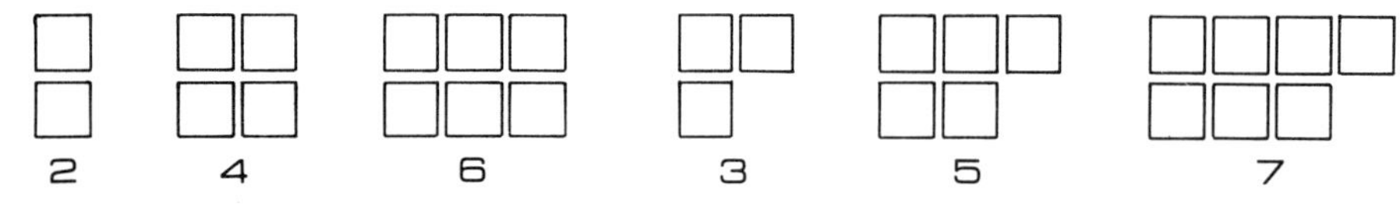

Fig. 13.2

♦ ♦ ♦ ♦ ♦ ♦ ♦ ♦

Pattern recognition involves many concepts, such as color and shape identification, direction, orientation, size, and number relationships. Children should use all these properties in identifying, extending, and creating patterns. Identifying the "cores" of patterns helps children become aware of the structures. For example, in some patterns the core repeats, whereas in others, the core grows.

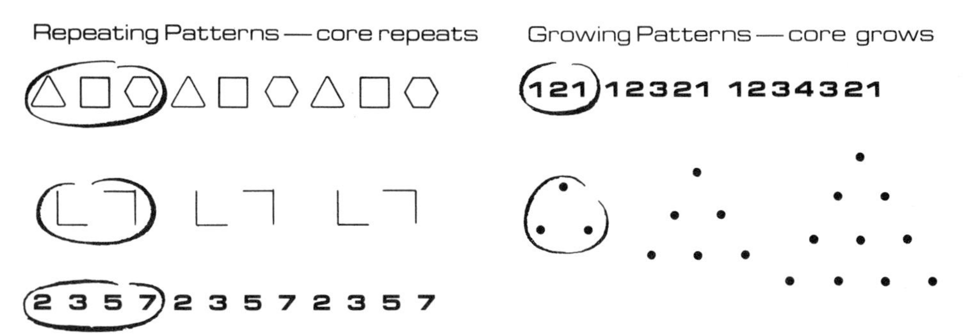

Representing a pattern both geometrically and numerically helps children recognize a variety of relationships in the pattern and make connections between arithmetic and geometry. See figure 13.3.

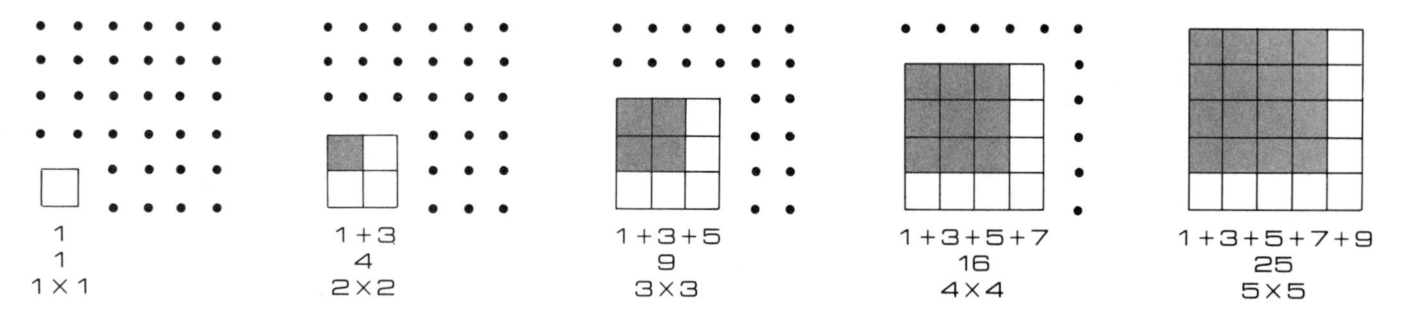

Fig. 13.3

Organizing data on a pattern in a table helps students identify its structure and describe it symbolically.

Pattern of Counters							
Element Number	1	2	3	4	5	...	□
Number of Counters	1	3	5	7	9	...	□+□−1

As they work on basic facts, children should be encouraged to look for patterns and relationships. The following sequence of numbers is an excellent example of such an activity:

9, 18, 27, 36, 45, 54, 63, 72, 81, 90

Children should recognize that each number is nine more than the number before it.

Each element should also be recognized as a multiple of 9 and represented as $9 \times n$. Replacing n with the numbers from 1 to 10 to generate the original set of numbers validates the symbolic representation of the pattern and reinforces multiplication facts. It also illustrates the concept of a variable.

1	2	3	4	5	6	7	8	9	10
11	12	13	14	15	16	17	18	19	20
21	22	23	24	25	26	27	28	29	30
31	32	33	34	35	36	37	38	39	40
41	42	43	44	45	46	47	48	49	50
51	52	53	54	55	56	57	58	59	60
61	62	63	64	65	66	67	68	69	70
71	72	73	74	75	76	77	78	79	80
81	82	83	84	85	86	87	88	89	90
91	92	93	94	95	96	97	98	99	100

Fig. 13.4

Coloring the elements on a hundred board represents the pattern in yet another way (see fig. 13.4). Such questions as the following help children to relate many mathematical ideas. Why do you think the numbers lie on a diagonal line? If you extended the hundred board to 200 and colored the next ten numbers in the pattern, where would they lie? Explain your thinking. How can you check your answer? What sequence of numbers starting with 9 would give a column of numbers on the hundred board?

Children should be encouraged to explore other less obvious patterns and relationships in the same sequence. For example, the units digits in succeeding elements decrease by 1, whereas the tens digits increase by 1, and the sum of the digits in each element is 9.

Using the constant function on a calculator, children can construct a table of input and output numbers and then express the relationship as an open sentence. See figure 13.5.

Input (I)	2	5	8	3	0	4	1	7
Output (O)	7	10	13	8	5	9	6	12

$O = I + 5$

Fig. 13.5

Graphing these sets of numbers helps children see number relationships in yet another format and is an informal extension to algebraic and geometric thinking.

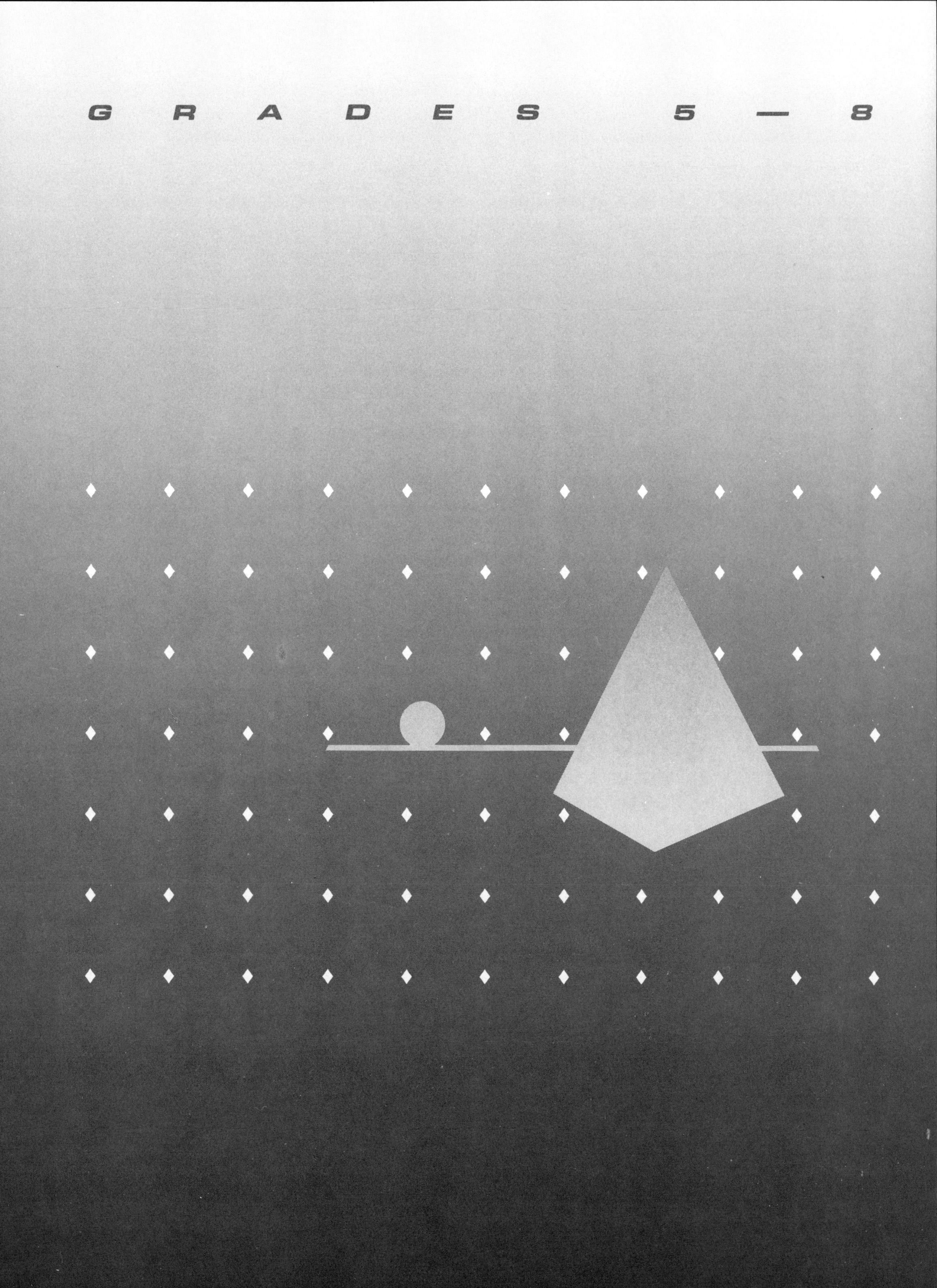
GRADES 5—8

♦ ♦ ♦ ♦ ♦ ♦ ♦ ♦

CURRICULUM STANDARDS FOR GRADES 5–8

OVERVIEW

This section presents thirteen curriculum standards for grades 5–8:

1. ***Mathematics as Problem Solving***
2. ***Mathematics as Communication***
3. ***Mathematics as Reasoning***
4. ***Mathematical Connections***
5. ***Number and Number Relationships***
6. ***Number Systems and Number Theory***
7. ***Computation and Estimation***
8. ***Patterns and Functions***
9. ***Algebra***
10. ***Statistics***
11. ***Probability***
12. ***Geometry***
13. ***Measurement***

The Need for Change

Mathematics is a useful, exciting, and creative area of study that can be appreciated and enjoyed by all students in grades 5–8. It helps them develop their ability to solve problems and reason logically. It offers to these curious, energetic students a way to explore and make sense of their world. However, many students view the current mathematics curriculum in grades 5–8 as irrelevant, dull, and routine. Instruction has emphasized computational facility at the expense of a broad, integrated view of mathematics and has reflected neither the vitality of the subject nor the characteristics of the students.

An ideal 5–8 mathematics curriculum would expand students' knowledge of numbers, computation, estimation, measurement, geometry, statistics, probability, patterns and functions, and the fundamental concepts of algebra. The need for this kind of broadened curriculum is acute. An examination of textbook series shows the repetition of topics, approach, and level of presentation in grade after grade. A comparison of the tables of contents shows little change over grades 5–8. It is even more disconcerting to realize that the very chapters that contain the most new material, such as probability, statistics, geometry, and prealgebra, are covered in the last half of the books—the sections most often skipped by teachers for lack of time. The result is an ineffective curriculum that rehashes material students already have seen. Such a curriculum promotes a negative image of mathematics and fails to give students an adequate background for secondary school mathematics.

These thirteen standards promote a broad curriculum for students in grades 5–8. Developing certain computational skills is important but constitutes only a part of this curriculum. Nevertheless, the existing curriculum in some schools prohibits many students from studying a broader curriculum until they have "mastered" basic computational skills. Shifting the focus to a broader curriculum is important for the following reasons:

1. Basic skills today and in the future mean far more than computational proficiency. Moreover, the calculator renders obsolete much of the complex paper-and-pencil proficiency traditionally emphasized in mathematics courses. Topics such as geometry, probability, statistics, and algebra have become increasingly more important and accessible to students through technology.

2. If students have not been successful in "mastering" basic computational skills in previous years, why should they be successful now, especially if the same methods that failed in the past are merely repeated? In fact, considering the effect of failure on students' attitudes, we might argue that further efforts toward mastering computational skills are counterproductive.

3. Many of the mathematics topics that are omitted actually can help students recognize the need for arithmetic concepts and skills and provide fresh settings for their use. For example, in probability, students have many opportunities to add and multiply fractions.

The vision articulated in the 5–8 standards is of a broad, concept-driven curriculum, one that reflects the full breadth of relevant mathematics and its interrelationships with technology. This vision is built on five overall curricular goals for students: learning to value mathematics, becoming confident in their ability, becoming a mathematical problem solver, learning to communicate mathematically, and learning to reason mathematically. The teaching of this curriculum should be related to the characteristics of middle school students and their current and future needs.

Features of the Mathematics Curriculum

The 5–8 curriculum should include the following features:

- Problem situations that establish the need for new ideas and motivate students should serve as the context for mathematics in grades 5–8. Although a specific idea might be forgotten, the context in which it is learned can be remembered and the idea re-created. In developing the problem situations, teachers should emphasize the application of mathematics to real-world problems as well as to other settings relevant to middle school students.

- Communication with and about mathematics and mathematical reasoning should permeate the 5–8 curriculum.

- A broad range of topics should be taught, including number concepts, computation, estimation, functions, algebra, statistics, probability, geometry, and measurement. Although each of these areas is valid mathematics in its own right, they should be taught as an integrated whole, not as isolated topics; the connections among them should be a prominent feature of the curriculum.

- Technology, including calculators, computers, and videos, should be used when appropriate. These devices and formats free students from tedious computations and allow them to concentrate on problem solving and other important content. They also give them new means to explore content. As paper-and-pencil computation becomes less important, the skills and understanding required to make proficient use of calculators and computers become more important.

Instruction

The standards are not intended to each constitute a chapter in a text or a particular unit of instruction; rather, learning activities should incorporate topics and ideas across standards. For example, an instructional activity might involve problem solving and use geometry, measurement, and computation. All mathematics should be studied in contexts that give the ideas and concepts meaning. Problems should arise from situations that are not always well formed. Students should have opportunities to formulate problems and questions that stem from their own interests.

Learning should engage students both intellectually and physically. They must become active learners, challenged to apply their prior knowledge and experience in new and increasingly more difficult situations. Instructional approaches should engage students in the process of learning rather than transmit information for them to receive. Middle grade students are especially responsive to hands-on activities in tactile, auditory, and visual instructional modes.

Classroom activities should provide students the opportunity to work both individually and in small- and large-group arrangements. The arrangement should be determined by the instructional goals as well as the nature of the activity. Individual work can help students develop confidence in their own ability to solve problems but should constitute only a portion of the middle school experience. Working in small groups provides students with opportunities to talk about ideas and listen to their peers, enables teachers to interact more closely with students, takes positive advantage of the social characteristics of the middle school student, and provides opportunities for students to exchange ideas and hence develops their ability to communicate and reason. Small-group work can involve collaborative or cooperative as well as independent work. Projects and small-group work can empower students to become more independent in their own learning. Whole-class discussions require students to synthesize, critique, and summarize strategies, ideas, or conjectures that are the products of individual and group work. These mathematical ideas can be expanded to, and integrated with, other subjects.

Materials

The 5–8 standards make the following assumptions about classroom materials:

- Every classroom will be equipped with ample sets of manipulative materials and supplies (e.g., spinners, cubes, tiles, geoboards, pattern blocks, scales, compasses, scissors, rulers, protractors, graph paper, grid-and-dot paper).

- Teachers and students will have access to appropriate resource materials from which to develop problems and ideas for explorations.

- All students will have a calculator with functions consistent with the tasks envisioned in this curriculum. Calculators should include the following features: algebraic logic including order of operations; computation in decimal and common fraction form; constant function for addition, subtraction, multiplication, and division; and memory, percent, square root, exponent, reciprocal, and +/− keys.

- Every classroom will have at least one computer available at all times for demonstrations and student use. Additional computers should be available for individual, small-group, and whole-class use.

Learner Characteristics

Implementation of the 5–8 standards should consider the unique characteristics of middle school students. As vast changes occur in their intellectual, psychological, social, and physical development, students in grades 5–8 begin to develop their abilities to think and reason more abstractly. Throughout this period, however, concrete experiences should continue to provide the means by which they construct knowledge. From these experiences they abstract more complex meanings and ideas. The use of language, both written and oral, helps students clarify their thinking and report their observations as they form and verify their mathematical ideas.

Students at this level can aptly be called "children in transition": they are restless, energetic, responsive to peer influence, and unsure about themselves. Self-consciousness is their hallmark, and curiosity about such questions as Who am I? How do I fit in? What do I enjoy doing? What do I want to be? is both their motivation and their nemesis. From this turmoil emerges an individual, with attitudes and patterns of thought taking shape.

In the transition to adulthood, middle school students are forming lifelong values and skills. The decisions students make about what they will study and how they will learn can dramatically affect their future. Failure to study mathematics can close the doors to vocational-technical schools, college majors, and careers—a loss of opportunity that happens most often to young women and minority students. Because many of the attitudes that affect these decisions are developed during the middle grades, it is crucial that conscious efforts be made to encourage all students, especially young women and minorities, to pursue mathematics. To this end, the curriculum must be interesting and relevant, must emphasize the usefulness of mathematics, and must foster a positive disposition toward mathematics.

Whenever possible, students' cultural backgrounds should be integrated into the learning experience. Black or Hispanic students, for example, may find the development of mathematical ideas in their cultures of great interest. Teachers must also be sensitive to the fact that students bring very different everyday experiences to the mathematics classroom. The way in which a student from an urban environment and a student from a suburban or rural environment interpret a problem situation can be very different. This is an important reason why communication is one of the overarching goals of these standards.

Students will perform better and learn more in a caring environment in which they feel free to explore mathematical ideas, ask questions, discuss their ideas, and make mistakes. By listening to students' ideas and encouraging them to listen to one another, one can establish an atmosphere of mutual respect. Teachers can foster this willingness to share by helping students explore a variety of ideas in reaching solutions and verifying their own thinking. This approach instills in students an understanding of the value of independent learning and judgment and discourages them from relying on an outside authority to tell them whether they are right or wrong.

Conclusion

Because the curriculum, activities, and mathematical knowledge envisioned in these standards are conceptually based, evaluation is not a simple or narrow task. The development of conceptual understanding is a long-term process; understanding is developed, elaborated, deepened, and made more nearly complete over time. Consequently, assessment must be an ongoing process. It should not be assumed that a single learning experience or assessment will provide a complete picture of students' intellectual growth. The Evaluation Standards (p. 189) offer many suggestions about this long-term assessment.

When interpreting these standards, developing curriculum, and integrating evaluation procedures, mathematics educators and others must realize that this broad, rich curriculum is intended to be available to *all* students. No student should be denied access to the study of one topic because he or she has yet to master another.

The current curriculum excludes many students from appreciating the useful, exciting, and creative aspects of mathematics. The 5–8 standards outline a curriculum that attempts to give all students the opportunity to appreciate the full power and beauty of mathematics and acquire the mathematical knowledge and intellectual tools necessary for its use in their lives.

The chart on the next page summarizes the major changes in emphasis for both the mathematical content and instruction in grades 5–8.

SUMMARY OF CHANGES IN CONTENT

INCREASED ATTENTION

PROBLEM SOLVING
- Pursuing open-ended problems and extended problem-solving projects
- Investigating and formulating questions from problem situations
- Representing situations verbally, numerically, graphically, geometrically, or symbolically

COMMUNICATION
- Discussing, writing, reading, and listening to mathematical ideas

REASONING
- Reasoning in spatial contexts
- Reasoning with proportions
- Reasoning from graphs
- Reasoning inductively and deductively

CONNECTIONS
- Connecting mathematics to other subjects and to the world outside the classroom
- Connecting topics within mathematics
- Applying mathematics

NUMBER/OPERATIONS/COMPUTATION
- Developing number sense
- Developing operation sense
- Creating algorithms and procedures
- Using estimation both in solving problems and in checking the reasonableness of results
- Exploring relationships among representations of, and operations on, whole numbers, fractions, decimals, integers, and rational numbers
- Developing an understanding of ratio, proportion, and percent

PATTERNS AND FUNCTIONS
- Identifying and using functional relationships
- Developing and using tables, graphs, and rules to describe situations
- Interpreting among different mathematical representations

ALGEBRA
- Developing an understanding of variables, expressions, and equations
- Using a variety of methods to solve linear equations and informally investigate inequalities and nonlinear equations

STATISTICS
- Using statistical methods to describe, analyze, evaluate, and make decisions

PROBABILITY
- Creating experimental and theoretical models of situations involving probabilities

GEOMETRY
- Developing an understanding of geometric objects and relationships
- Using geometry in solving problems

continued on p. 72

AND EMPHASIS IN 5–8 MATHEMATICS

DECREASED ATTENTION

PROBLEM SOLVING
- Practicing routine, one-step problems
- Practicing problems categorized by types (e.g., coin problems, age problems)

COMMUNICATION
- Doing fill-in-the-blank worksheets
- Answering questions that require only yes, no, or a number as responses

REASONING
- Relying on outside authority (teacher or an answer key)

CONNECTIONS
- Learning isolated topics
- Developing skills out of context

NUMBER/OPERATIONS/COMPUTATION
- Memorizing rules and algorithms
- Practicing tedious paper-and-pencil computations
- Finding exact forms of answers
- Memorizing procedures, such as cross-multiplication, without understanding
- Practicing rounding numbers out of context

PATTERNS AND FUNCTIONS
- Topics seldom in the current curriculum

ALGEBRA
- Manipulating symbols
- Memorizing procedures and drilling on equation solving

STATISTICS
- Memorizing formulas

PROBABILITY
- Memorizing formulas

GEOMETRY
- Memorizing geometric vocabulary
- Memorizing facts and relationships

continued on p. 73

SUMMARY OF CHANGES—continued

INCREASED ATTENTION

MEASUREMENT

- Estimating and using measurement to solve problems

INSTRUCTIONAL PRACTICES

- Actively involving students individually and in groups in exploring, conjecturing, analyzing, and applying mathematics in both a mathematical and a real-world context
- Using appropriate technology for computation and exploration
- Using concrete materials
- Being a facilitator of learning
- Assessing learning as an integral part of instruction

DECREASED ATTENTION

MEASUREMENT

- Memorizing and manipulating formulas
- Converting within and between measurement systems

INSTRUCTIONAL PRACTICES

- Teaching computations out of context
- Drilling on paper-and-pencil algorithms
- Teaching topics in isolation
- Stressing memorization
- Being the dispenser of knowledge
- Testing for the sole purpose of assigning grades

STANDARD 1: MATHEMATICS AS PROBLEM SOLVING

In grades 5–8, the mathematics curriculum should include numerous and varied experiences with problem solving as a method of inquiry and application so that students can—

- ***use problem-solving approaches to investigate and understand mathematical content;***
- ***formulate problems from situations within and outside mathematics;***
- ***develop and apply a variety of strategies to solve problems, with emphasis on multistep and nonroutine problems;***
- ***verify and interpret results with respect to the original problem situation;***
- ***generalize solutions and strategies to new problem situations;***
- ***acquire confidence in using mathematics meaningfully.***

Focus

> To solve a problem is to find a way where no way is known off-hand, to find a way out of a difficulty, to find a way around an obstacle, to attain a desired end, that is not immediately attainable, by appropriate means. (G. Polya in Krulik and Reys 1980, p. 1)

Problem solving is the process by which students experience the power and usefulness of mathematics in the world around them. It is also a method of inquiry and application, interwoven throughout the *Standards* to provide a consistent context for learning and applying mathematics. Problem situations can establish a "need to know" and foster the motivation for the development of concepts.

In grades 5–8, the curriculum should take advantage of the expanding mathematical capabilities of middle school students to include more complex problem situations involving topics such as probability, statistics, geometry, and rational numbers. Situations and approaches should build on and extend the mathematical language students are acquiring and help them to develop a variety of problem-solving strategies and approaches. Although concrete and empirical situations remain a focus throughout these grades, a balance should be struck between problems that apply mathematics to the real world and problems that arise from the investigation of mathematical ideas. Finally, the mathematics curriculum should engage students in some problems that demand extended effort to solve. Some might be group projects that require students to use available technology and to engage in cooperative problem solving and discussion. For grades 5–8 an important criterion of problems is that they be interesting to students.

Computers and calculators are powerful problem-solving tools. The power to compute rapidly, to graph a relationship instantly, and to systematically change one variable and observe what happens to other related variables can help students become independent doers of mathematics.

The curriculum must give students opportunities to solve problems that require them to work cooperatively, to use technology, to address relevant and interesting mathematical ideas, and to experience the power and usefulness of mathematics.

Discussion

The nonroutine problem situations envisioned in these standards are much broader in scope and substance than isolated puzzle problems. They are also very different from traditional word problems, which provide contexts for using particular formulas or algorithms but do not offer opportunities for true problem solving. Real-world problems are not ready-made exercises with easily processed procedures and numbers. Situations that allow students to experience problems with "messy" numbers or too much or not enough information or that have multiple solutions, each with different consequences, will better prepare them to solve problems they are likely to encounter in their daily lives.

The exploration of problem situations can provide a context in which students further their knowledge about the interrelationships of mathematical ideas, for example:

Maria used her calculator to explore this problem: Select five digits. Use the five digits to form a two-digit and a three-digit number so that their product is the largest possible. Then find the arrangement that gives the smallest product.

Students can be encouraged to generalize their solutions to any five digits and to any number of digits. This problem helps students deepen their understanding of place value, multiplication, and number sense.

Students should model many problems concretely, gather and organize data in tables, identify patterns, graph data, use calculators to simplify computations, and use computers to assist in generating and analyzing information. The power of computers to store, generate, and depict—in various ways—vast quantities of information makes them a valuable source of interesting problems. Data bases and computer programs can engage students in posing and solving problems. Students sample data, analyze and make predictions on the basis of their samples, make conjectures, discuss and validate their conclusions, and prepare arguments to convince others of their conclusions. Students also should experience problem situations rich in opportunities to formulate and define problems, determine the information required, decide on methods for obtaining this information, and determine the limits of acceptable solutions. The following is one example of such an open-ended problem situation:

The teacher demonstrates a pendulum constructed from string and a weight. Students work in small groups to construct a pendulum, investigate how it functions, and formulate questions that arise. These questions might include, How long does it take to make one complete cycle? How does the length of the string affect the cycle? How does the weight affect the cycle? How does the height from which the pendulum begins its swing affect the cycle? How long does it take before it comes to rest?

Groups might first share their questions with the whole class and then, in small groups, decide which questions they wish to investigate. Such situations allow students to formulate questions based on their own interests.

Students should frequently work together in small groups to solve problems. They can discuss strategies and solutions, ask questions, examine consequences and alternatives, and reflect on the process and how it re-

♦ ♦ ♦ ♦ ♦ ♦ ♦ ♦

lates to prior problems. Students must verify results, interpret solutions, and question whether a solution makes sense. They should verify their own thinking rather than depend on the teacher to tell them whether they are right or wrong. Such experiences develop students' confidence in using mathematics.

Instruction should also help students develop their ability to understand and apply a variety of strategies (e.g., guess and check, make a table, look for patterns). These strategies should be explored in the context of solving problems. Through group and classroom discussions, students can examine a variety of approaches and learn to evaluate appropriate strategies for a given situation. The instructional goal is that students will build an increasing repertoire of strategies, approaches, and familiar problems; it is the problem-solving process that is most important, not just the answer. The following problem illustrates how students might share their approaches in solving problems:

How many handshakes will occur at a party if every one of the 15 guests shakes hands with each of the others?

Some students will choose to act out the problem. Some might draw a picture of a simpler case to approximate the situation (fig. 1.1). Other students might start with a simpler problem and look for a pattern (fig. 1.2):

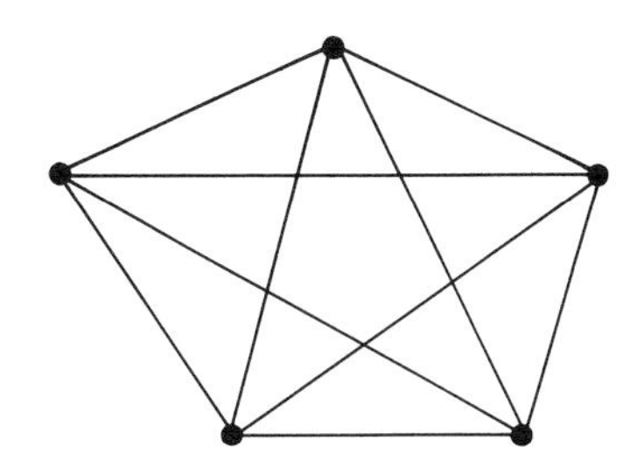

***Fig. 1.1.* Drawing a picture of a simpler code**

# people	1	2	3	4	5	6	...	15
# handshakes	0	1	3	6	10	15	...	105

***Fig. 1.2.* Looking for patterns**

Class discussions in which students share their approaches enrich and expand their repertoire of strategies for solving problems.

In addition to cooperative effort, real-world problems often require a substantial investment of time. Students should be encouraged to explore some problems as extended projects that can be worked on for hours, days, or longer. For example, students concerned about the traffic congestion at an intersection near their school decided to design a method to study the situation. Their study required that they answer such questions as what constituted "traffic," when to count it, how to record data, what the data meant, how to alleviate the congestion, who was responsible for dealing with such problems, and how to construct a convincing argument that the situation should be remedied. After several weeks of study, the results were presented to the city council, which accepted the students' recommendation and installed a traffic signal.

Not all problems require a real-world setting. Indeed, middle school students often are intrigued by story settings or those arising from mathematics itself. For example, which number from 1 to 100 has the most factors? What matters is that students experience mathematics in situations in which they come to view it as personally empowering. For example, the ability to see and use similar triangles to solve a nonroutine indirect measurement problem is empowering in two ways. It gives students confidence in their ability and perception to solve such problems in their own lives. Further, they see that their mathematical thinking contributes to their mastery of other mathematical ideas, such as their understanding of the relationship between ratio and fraction concepts.

The 5–8 standards contain numerous examples of how problem situations can serve as a context for exploring mathematical ideas. Through these situations, students have opportunities to investigate problems, apply their knowledge and skills across a wide range of situations, and develop an appreciation for the power and beauty of mathematics.

STANDARD 2: MATHEMATICS AS COMMUNICATION

In grades 5–8, the study of mathematics should include opportunities to communicate so that students can—

- ***model situations using oral, written, concrete, pictorial, graphical, and algebraic methods;***
- ***reflect on and clarify their own thinking about mathematical ideas and situations;***
- ***develop common understandings of mathematical ideas, including the role of definitions;***
- ***use the skills of reading, listening, and viewing to interpret and evaluate mathematical ideas;***
- ***discuss mathematical ideas and make conjectures and convincing arguments;***
- ***appreciate the value of mathematical notation and its role in the development of mathematical ideas.***

Focus

The use of mathematics in other disciplines has increased dramatically, largely because of its power to represent and communicate ideas concisely. Society's increasing use of technology requires that students learn both to communicate with computers and to make use of their own individual power as a medium of communication. The ability to read, write, listen, think creatively, and communicate about problems will develop and deepen students' understanding of mathematics.

Middle school students should have many opportunities to use language to communicate their mathematical ideas. The communication process requires students to reach agreement about the meanings of words and to recognize the crucial importance of commonly shared definitions. Opportunities to explain, conjecture, and defend one's ideas orally and in writing can stimulate deeper understandings of concepts and principles. It is essential that mathematical concepts be firmly attached to the symbols that represent them; the need for symbolic representation arises out of the exploration of these concepts. In the process of discussing mathematical concepts and symbols, students become aware of the connections between them. Unless students frequently and explicitly discuss relationships between concepts and symbols, they are likely to view symbols as disparate objects to be memorized.

As students progress from grade 5 to grade 8, their ability to reason abstractly matures greatly. Concurrent with this enhanced ability to abstract common elements from situations, to conjecture, and to generalize—in short, to *do* mathematics—should come an increasing sophistication in the ability to *communicate* mathematics. But this development cannot occur without deliberate and careful acquisition of the language of mathematics.

Discussion

Communication involves the ability to read and write mathematics and to interpret meanings and ideas. Writing and talking about their thinking clarifies students' ideas and gives the teacher valuable information from which to make instructional decisions. Emphasizing communication in a mathematics class helps shift the classroom from an environment in which students are totally dependent on the teacher to one in which students assume more responsibility for validating their own thinking.

Teachers foster communication in mathematics by asking questions or posing problem situations that actively engage students, including situations that encourage students to create problems themselves. Small-group work, large-group discussions, and the presentation of individual and group reports—both written and oral—create an environment in which students can practice and refine their growing ability to communicate mathematical thought processes and strategies. Small groups provide a forum in which students ask questions, discuss ideas, make mistakes, learn to listen to others' ideas, offer constructive criticism, and summarize their discoveries in writing. Whole-class discussions enable students to pool and evaluate ideas, record data, share solution strategies, summarize collected data, invent notations, hypothesize, and construct simple arguments. For example, a teacher might present the class with the following situation:

A national magazine surveyed teenagers to determine the number of hours of TV they watched each day. How many hours do you think the magazine reported?

Students can discuss their predictions in small groups, write summaries of their group work or of their own ideas, share their predictions with the class, discuss their reasoning, and compare their predictions with the magazine's report. This exercise encourages students to evaluate the magazine report, discuss appropriate survey techniques, design and conduct their own survey and compare it with the national survey, prepare a written report, compare results from different groups, and evaluate their findings. As students refine their communication skills, they gain confidence in their ability to build convincing mathematical arguments.

Students' development of mathematical concepts and the language and symbols needed to describe and represent concepts is enhanced by carefully planned and orchestrated sequences of concrete situations that build a need for, and give meaning to, the symbols. These interconnections must be taught directly through varied examples and verbalization. For example, many students are intrigued by number tricks, such as that in figure 2.1: "Think of a number. Add 5 to it. Multiply the result by 2. Subtract 4. Divide by 2. Subtract the number you first thought of. I bet I can read your mind—your answer is 3." Modeling the situation with concrete materials, such as tiles and beans, builds one notion of variable and can lead to algebraic notation.

This example illustrates the role of written symbols in representing ideas, a concept that is developed throughout the middle school years. Students learn to use precise language in conjunction with the special symbol systems of mathematics, such as algebraic notation.

the number thought of:	□	n
add five:	□ ○ ○ ○ ○ ○	$n+5$
multiply by two:	□ □ ○ ○ ○ ○ ○ ○ ○ ○ ○ ○	$2n+10$
subtract four:	□ □ ○ ○ ○ ○ ○ ○	$2n+6$
divide by two:	□ ○ ○ ○	$n+3$
subtract the number thought of:	○ ○ ○	3

***Fig. 2.1.* Number trick**

This standard is firmly tied to problem solving and reasoning. As students' mathematical language develops, so does their ability to reason about and solve problems. Moreover, problem-solving situations provide a setting for the development and extension of communication skills and reasoning ability. The following problem illustrates how students might share their approaches in solving problems:

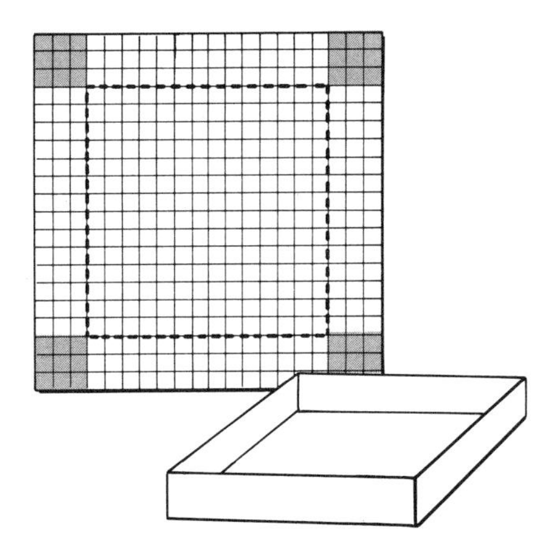

Fig. 2.2. **Building a grid-paper box**

The class is divided into small groups. Each group is given square pieces of grid paper and asked to make boxes by cutting out pieces from the corners. Each group is given 20 × 20 grid paper. See figure 2.2. Students cut and fold the paper to make boxes sized 18 × 18 × 1, 16 × 16 × 2, . . ., 2 × 2 × 8. They are challenged to find a box that holds the maximum volume and to convince someone else that they have found the maximum. Groups are encouraged also to explore other grid sizes, such as 19 × 19 or 24 × 24.

Some groups might decide not to limit themselves to boxes that are cut on the lines. Others might make a graph of the volume as compared to the height of the box. One group might decide to see what happens when they use the scraps left over from the corners. In a class discussion, students share their explicit findings, from which they eventually extrapolate generalizations. Students recognize that their solutions depend on the way in which they define the problem. These kinds of explorations provide opportunities for students to write about their ideas and the generalizations they have made.

Teachers' questioning techniques should help students construct connections among concepts, procedures, and approaches. Questions that limit answers to recitation of a single number, a simple yes or no, or a memorized procedure do not teach students the communication skills they will need. Consider instead the following exercises:

Give examples of a rectangle with four congruent sides; a parallelogram with four right angles; a trapezoid with two equal angles; a number between 1/3 and 1/2; a number with a repeating decimal representation; a jacket for a cube; an equation for a line that passes through the point (−1, 2).

These more open-ended problems can have several correct answers and can promote opportunities for students to write about their ideas, discuss interpretations, and expand their understandings.

An interchange occurs between common and mathematical language. Mathematical language builds on the existing structure and logic of common language and connects students' experiences and language to the mathematical world. Terms whose meanings change from one language to another must be addressed straightforwardly. For example, the use of such terms as *improper fraction* and *right angle* as mathematical descriptions can be misleading to students, who relate them to the common meanings of the words *improper* and *right*.

The NCTM position statement "Mathematics for Language Minority Students" (NCTM 1987) states that "cultural background or difficulties with the English language must not exclude any student from full participation in the school's mathematics program." Students whose primary language is not standard English may require special support to facilitate their learning of mathematics.

The ability to read, listen, think creatively, and communicate about problem situations, mathematical representations, and the validation of solutions will help students to develop and deepen their understanding of mathematics.

STANDARD 3: MATHEMATICS AS REASONING

In grades 5–8, reasoning shall permeate the mathematics curriculum so that students can—

- ***recognize and apply deductive and inductive reasoning;***
- ***understand and apply reasoning processes, with special attention to spatial reasoning and reasoning with proportions and graphs;***
- ***make and evaluate mathematical conjectures and arguments;***
- ***validate their own thinking;***
- ***appreciate the pervasive use and power of reasoning as a part of mathematics.***

Focus

Reasoning is fundamental to the knowing and doing of mathematics. Although most disciplines have standards of evaluation by which new theories or discoveries are judged, nowhere are these standards as explicit and well formulated as they are in mathematics. Conjecturing and demonstrating the logical validity of conjectures are the essence of the creative act of doing mathematics. To give more students access to mathematics as a powerful way of making sense of the world, it is essential that an emphasis on reasoning pervade all mathematical activity. Students need a great deal of time and many experiences to develop their ability to construct valid arguments in problem settings and evaluate the arguments of others.

The development of logical reasoning is tied to the intellectual and verbal development of students. Through grades 5–8, students' reasoning abilities change. Whereas most fifth graders still are concrete thinkers who depend on a physical or concrete context for perceiving regularities and relationships, many eighth-grade students are capable of more formal reasoning and abstraction. Even the most advanced students at the 5–8 level, however, might use concrete materials to support their reasoning; this is especially true for spatial reasoning. The 5–8 mathematics curriculum should pay special attention to the development of student's abilities to use proportional and spatial reasoning and to reason from graphs.

Technology can foster environments in which students' growing curiosity can lead to rich mathematical invention. In these environments, the control of exploring mathematical ideas is turned over to students. Both inductive and deductive reasoning come into play as students make conjectures and seek to explain why they are valid. Whether encouraged by technology or by challenging mathematical situations posed in the classroom, this freedom to explore, conjecture, validate, and to convince others is critical to the development of mathematical reasoning in the middle grades.

Discussion

The seeds of logical thinking are planted as students learn to describe objects or processes accurately and to elaborate their properties, similarities, differences, and relationships. Students should be encouraged to explain their reasoning in their own words. Listening to their peers and their teacher describe other strategies helps students refine their thoughts and the language they use to express their thoughts. Such questions as the following should abound in the mathematics classes: Why? What if . . .? Can you give an example of . . .? Can you find a counterexample? Do you see a pattern? Is this always true? Sometimes true? Never true? How do you know? Such questions prompt students to validate and value their own thinking.

Identifying patterns is a powerful problem-solving strategy. It is also the essence of inductive reasoning. As students explore problem situations appropriate to their grade level, they can often consider or generate a set of specific instances, organize them, and look for a pattern. These, in turn, can lead to conjectures about the problem. Students should be encouraged to validate these conjectures by constructing supporting arguments, which can be at many levels of sophistication.

Students at these grade levels should be exposed to problem situations that are challenging but within reach. For example, students can be asked to explore the numbers that occur between twin primes for primes greater than 3. They might first look for twin primes to find examples and then make, test, and validate conjectures.

5 **6** 7; 11 **12** 13; 17 **18** 19; 29 **30** 31

What do 6, 12, 18, and 30 have in common? Is this true for all twin primes? Why or why not? This problem presents an excellent opportunity for a class to use a computer to generate lists of primes.

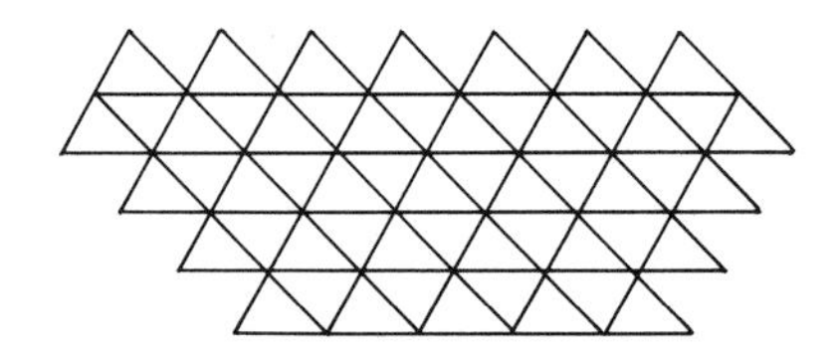

***Fig. 3.1.* Tiling method**

Students can be introduced to many kinds of mathematical reasoning. To help them recognize one aspect of the beauty of mathematics, groups of students can each be asked to cut out a triangle of their choice and then see whether they can "tile" the plane using copies of their own triangle. Less sophisticated students might say, "It works!" on the basis of a single instance. Others might look at the group's examples and see that the pattern seems to work for every triangle; students eventually might reason from the angle-sum property of triangles that the tiling method always works. See figure 3.1.

Students can use reasoning to illustrate when something never works. For example, students can be given a particular collection of numbers (3, 6, 12, 15, 21, 27, 42, 51) and be asked to find a set of these numbers that sums to 100. Once it is clear that no one will succeed, students can be challenged to reason why such a sum is impossible: Any sum of multiples of 3 is a multiple of 3, so the sum cannot be 100.

Students should also encounter situations in which reasoning from a counterexample is useful: Suppose one has two numbers that divide 72. Does their product also divide 72? (2 divides 72; 3 divides 72; 2 × 3 divides 72.) Is this always true? A counterexample, such as 4 divides 72 and 8 divides 72, shows that the product does not always divide 72.

The ability to reason proportionally develops in students throughout grades 5–8. It is of such great importance that it merits whatever time and effort must be expended to assure its careful development. Students need to see many problem situations that can be modeled and then solved through proportional reasoning. Such problems can range from simple to complex, as illustrated by the three problems below:

Students observe that their classroom has 16 windowpanes. If every room has 16 panes, how many windowpanes are in a 20-room school?

♦ ♦ ♦ ♦ ♦ ♦ ♦ ♦

A shop sells special cookies for $1 each, or 10 for $9. Tom wishes to buy 30 cookies. How much should he pay? (Vergnaud 1988)

A group of 8 people are going camping for 3 days and need to carry their own water. They read in a guide book that 12.5 liters are needed for a party of 5 persons for 1 day. How much water should they carry? (Vergnaud 1988)

Geometric as well as number situations should be created. For example, similarity of figures and scaling—in fact, all scale-model-to-real-object problems—provide appropriate settings for proportional reasoning.

If given opportunities to reason from graphs about interesting situations, students can develop an appreciation for the problem-solving potential of making, using, and talking about graphs. The following example (Swan 1985) offers a flavor of the potential of graphical representations as tools for reasoning.

Students are given a carefully drawn picture of a roller-coaster track (fig. 3.2).

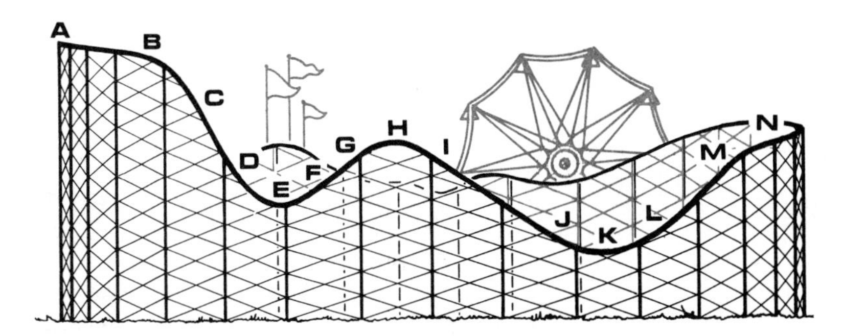

***Fig. 3.2.* Roller coaster**

The challenge is to sketch a graph (with no numbers) to represent the speed of the roller coaster versus its position on the track.

Now, to reverse the problem, students are given a part of the graph of speed versus position for another roller coaster (fig. 3.3). The question becomes,

What does the roller-coaster track look like?

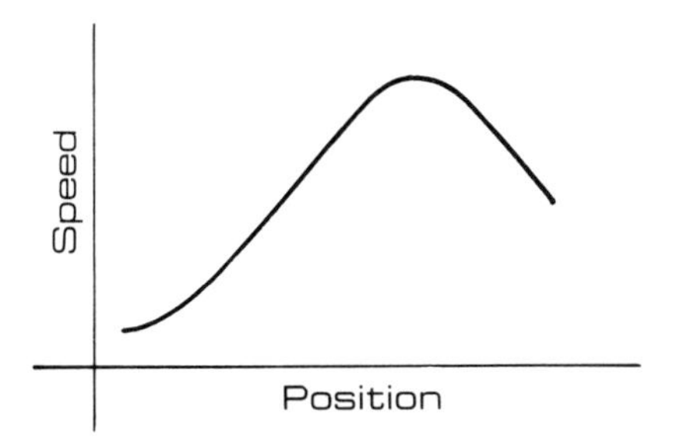

***Fig. 3.3* Roller-coaster graph**

Investigating graphical representations and their relationships to algebraic representations can give students a real sense of the dynamic relationship between the variables. Such problem settings also allow students to reason directly to, and hypothetically from, graphs.

Students can develop their spatial reasoning abilities in a variety of interesting settings. They can gather a collection of small objects, such as spools, golf balls, small footballs, small cans and bottles, and foam cups, and then try to draw what they think the shape of the shadow of each object might look like. Students can then test their conjectures by using the overhead projector to cast shadows; they can also be asked to identify the object solely on the appearance of its shadow. Relating the size of the shadow to the size of the object also allows the use of proportional thinking.

Reasoning about what is happening and why should be a constant part of the study of mathematics. Students in grades 5–8 should explore mathematical reasoning through problem situations that are appropriate to their ages and interest.

STANDARD 4: MATHEMATICAL CONNECTIONS

In grades 5–8, the mathematics curriculum should include the investigation of mathematical connections so that students can—

- ***see mathematics as an integrated whole;***
- ***explore problems and describe results using graphical, numerical, physical, algebraic, and verbal mathematical models or representations;***
- ***use a mathematical idea to further their understanding of other mathematical ideas;***
- ***apply mathematical thinking and modeling to solve problems that arise in other disciplines, such as art, music, psychology, science, and business;***
- ***value the role of mathematics in our culture and society.***

Focus

For many students, mathematics in the middle grades has far too often simply repeated or extended much of the computational work covered in the earlier grades. The intent of this standard is to help students broaden their perspective, to view mathematics as an integrated whole rather than as an isolated set of topics, and to acknowledge its relevance and usefulness both in and out of school. Mathematics instruction at the 5–8 level should prepare students for expanded and deeper study in high school through exploration of the interconnections among mathematical ideas.

Students should have many opportunities to observe the interaction of mathematics with other school subjects and with everyday society. To accomplish this, mathematics teachers must seek and gain the active participation of teachers of other disciplines in exploring mathematical ideas through problems that arise in their classes. This integration of mathematics into contexts that give its symbols and processes practical meaning is an overarching goal of all the standards. It allows students to see how one mathematical idea can help them understand others, and it illustrates the subject's usefulness in solving problems, describing and modeling real-world phenomena, and communicating complex thoughts and information in a concise and precise manner. Different representations of problems serve as different lenses through which students interpret the problems and the solutions. If students are to become mathematically powerful, they must be flexible enough to approach situations in a variety of ways and recognize the relationships among different points of view.

This standard precedes those that discuss specific mathematical content in order to stress the importance of viewing the standards as an integrated whole, not as a list of content areas. If the standards are interpreted as a listing of topics to be covered sequentially, it is likely that in most classrooms there will be insufficient time to cover them all. Instead, implementation of the standards should be organized in such a way that several goals will be addressed simultaneously.

Discussion

Connections among various mathematical topics can be drawn in many ways. When a student makes and describes the translation in geometry "20 to the right" and follows it with a second translation, "45 to the left," the result is fundamentally identical to adding integers. Various interpretations of fractions can illustrate connections to measurement, ratios, and ideas in algebra. As students study one topic, relationships to other topics can be highlighted and applied. Connections also can emerge as students do mathematics. The development and exploration of patterns in Pascal's triangle, for example, can be used to illustrate relationships among counting, exponents, algebra, geometric patterns, probability, and number theory. Although these connections should not be made in a formal way at the 5–8 level, teachers can foster an informal familiarity with them through problem solving: "Can you find the triangular numbers in the Pascal triangle? Why do they appear? Does Pascal's triangle have anything to do with the probability of getting two heads in three flips of a coin? Why?"

This persistent attention to recognizing and drawing connections among topics will instill in students an expectation that the ideas they learn are useful in solving other problems and exploring other mathematical concepts. For example, a knowledge of area can help them in understanding the operations on fractions, representing statistical data, solving proportion problems, finding factors and probabilities, and exploring the meanings of algebraic expressions. As they learn new ideas or solve new problems, students enrich their own thought processes and skills by drawing on previously developed ideas; this ability to integrate ideas and concepts fosters students' confidence in their own thinking as well as in their skills of communication. Curriculum materials can foster an attitude in students that will encourage them to look for connections, but teachers must also look for opportunities to help students make mathematical connections.

Technology is useful for identifying connections. For example, in investigating computer-generated spirals of line segments, students might find that the size of the initial angle, which they enter into the computer, is related to the "shape" of the spiral; for example, an angle of 90 degrees generates a "square" spiral. Because the computer allows students to enter countless values and immediately see the resulting geometric shape, they might find it both interesting and rewarding to investigate interrelationships between number and geometry: An angle of 72 degrees gives a five-sided spiral, and $2 \times 72 = 144$ degrees gives a five-sided star.

Varied problem settings are a means by which students can highlight and build mathematical connections. A lesson on measurement can be an occasion for students to formulate and solve problems while exploring geometric, measurement, and algebraic ideas. Consider the following situation with pattern blocks (fig. 4.1):

Describe the pattern. What questions can we ask about the pattern?

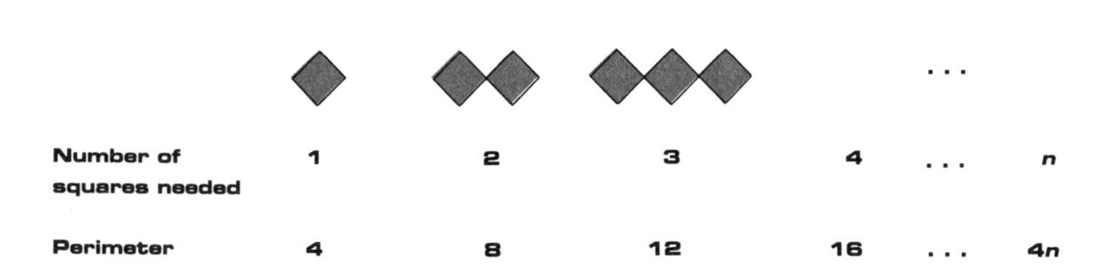

***Fig. 4.1.* Situation A**

This situation and that in figure 4.2 allow students to make conjectures and to convince others that their ideas are valid. In addition, questions about the area (number of squares) or the perimeter of the tenth or the nth term in the patterns are likely to arise.

Fig. 4.2. Situation B

Number of triangles needed	**1**	**4**	**9**	**16**	**. . .**	n^2
Perimeter	**3**	**6**	**9**	**12**	**. . .**	**3n**

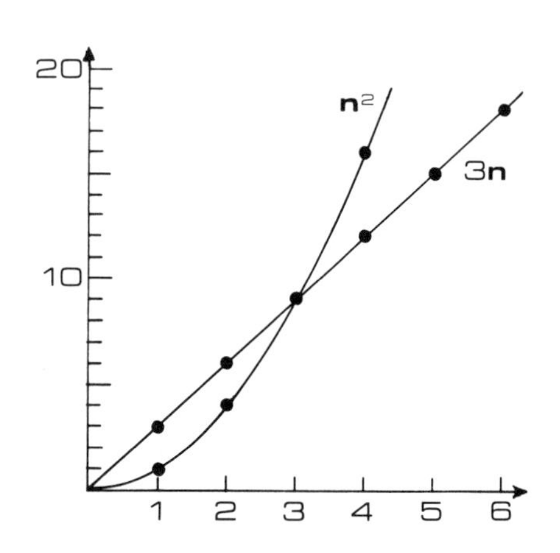

Fig. 4.3. Pattern growth

We can represent the changing area (the number of triangles) or perimeter by a number sequence, a verbal description, or an algebraic rule. If we want to compare the growth of the perimeter and the number of triangles needed, we can make a graph on the same axes. The graph for situation B is shown in figure 4.3.

This type of interconnected pattern work can lead to surprises for both teachers and students. Students often identify patterns very different from those the teacher had anticipated. This creative aspect of making sense of mathematics is a real confidence builder for students. Such situations also allow the class to explore various ways to model the relationships mathematically and determine how these models are alike and how each can highlight a different aspect of the problem. The more rapid growth of the area in situation B can be seen from the numbers, but the graph shows the change in a more dynamic way. One of the most important connections students in grades 5–8 should understand is this relationship between data in tables, algebraic generalizations, and graphical representations.

Many opportunities to show the connections between mathematics and other disciplines are missed in school. Mathematics arises not only in science but in other subjects as well. In social studies, for example, the study of maps is an excellent time to also study scaling and its relation to the concepts of similarity, ratio, and proportion. A topic such as measurement has implications for social studies, science, home economics, industrial technology, and physical education and is increasingly important to teachers of these subjects.

"Connected" mathematics should not be disconnected from students' daily lives. For example, although the "handshake" problem can be used to show connections between triangular numbers and the diagonals of a polygon, classroom discussion might focus on reasons why airlines use "hub cities" to map routes among the cities they serve. As students in grades 5–8 become aware of the world around them, probability and statistics become increasingly important connections between the real world and the mathematics classroom. Weather forecasting, scientific experiments, advertising claims, chance events, and economic trends are but a few of the areas in which students can investigate the role of mathematics in our society. Statistics offer students insights into problems of social equity. Perspective, proportion, and the golden ratio are ways of learning mathematics in the context of art and design. Whatever the context, a vital role of mathematics education is to instill in students an attitude of inquiry and investigation and a sensitivity to the many interrelationships between formal mathematics and the real world.

STANDARD 5: NUMBER AND NUMBER RELATIONSHIPS

In grades 5–8, the mathematics curriculum should include the continued development of number and number relationships so that students can—

- ***understand, represent, and use numbers in a variety of equivalent forms (integer, fraction, decimal, percent, exponential, and scientific notation) in real-world and mathematical problem situations;***
- ***develop number sense for whole numbers, fractions, decimals, integers, and rational numbers;***
- ***understand and apply ratios, proportions, and percents in a wide variety of situations;***
- ***investigate relationships among fractions, decimals, and percents;***
- ***represent numerical relationships in one- and two-dimensional graphs.***

Focus

The use of concise symbols and language to represent numbers is a significant historical and practical development. In the middle school years, students come to recognize that numbers have multiple representations, so the development of concepts for fractions, ratios, decimals, and percents and the idea of multiple representations of these numbers need special attention and emphasis. The ability to generate, read, use, and appreciate multiple representations of the same quantity is a critical step in learning to understand and do mathematics.

As students progress through middle school, they build their knowledge of rational numbers as important both in their own right and as a foundation for rational forms in algebra. Ratio, proportion, and percent are introduced and developed in grades 5–8. In exploring these topics, students have many opportunities to develop their ability to reason proportionally.

To provide students with a lasting sense of number and number relationships, learning should be grounded in experience related to aspects of everyday life or to the use of concrete materials designed to reflect underlying mathematical ideas. Students should encounter number lines, area models, and graphs as well as representations of numbers that appear on calculators and computers (e.g., forms of scientific notation). Students should learn to identify equivalent forms of a number and understand why a particular representation is useful in a given setting.

Discussion

Understanding multiple representations for numbers is a crucial precursor to solving many of the problems students encounter. Toward this end, students can represent fractions, decimals, and percents in a variety of meaningful situations, thereby learning to move flexibly among concrete, pictorial, and abstract representations. Students should understand these numbers and their representations, the relationships among them, and the advantages and disadvantages of each.

Teachers should strive to make this process consistently positive; too often, students are taught that $\frac{2}{4} = \frac{1}{2}$, only to be informed later that $\frac{2}{4}$ is a "wrong answer" when the "correct" answer is $\frac{1}{2}$. Discussing the appropriateness of certain representations in a given situation, such as the fact that it is better to write "68/100 dollars" on a check than reduce to "17/25 dollars," helps students recognize that there is no single, uniform way to represent a fraction but that the "best" way depends largely on the situation. Students learn, for example, that $\frac{15}{100}$, $\frac{3}{20}$, 0.15, and 15% are all representations of the same number, appropriate for a fraction of a dollar on a bank check, the probability of winning a game, the tax on a purchase of $2.98, and a discount, respectively. Similarly, they learn that +8, $\frac{8}{1}$, and 8.0 are all appropriate representations of the same number, depending on whether they are subtracting integers, adding fractions, or labeling a coordinate axis with rational numbers.

Exponents and scientific notation give further examples of the elegance and complexities of concise notation. Numbers like 2^{30} in the denominator of a fraction in a probability problem, 1.06^{10} as a factor in a population growth problem, 9.3×10^{-8} in a science problem, and $3.46 million in a newspaper article are approximate representations, each of which is best suited to its own context. Calculators and computers can display forms like 7.23471 07, thus requiring that students learn other forms of scientific notation. Students also should recognize the difficulties inherent in various representations, such as the expression of $\frac{1}{3}$ as a percent, 5.7×10^{-9} as an ordinary decimal, or 5.999999999 on a calculator as 6. They also need to grasp some sense of the "infinite" quality of decimals.

Area models are especially helpful in visualizing numerical ideas from a geometric point of view. For example, area models can be used to show that $\frac{8}{12}$ is equivalent to $\frac{2}{3}$, that $1.2 \times 1.3 = 1.56$, and that 80% of 20 is 16. See figure 5.1.

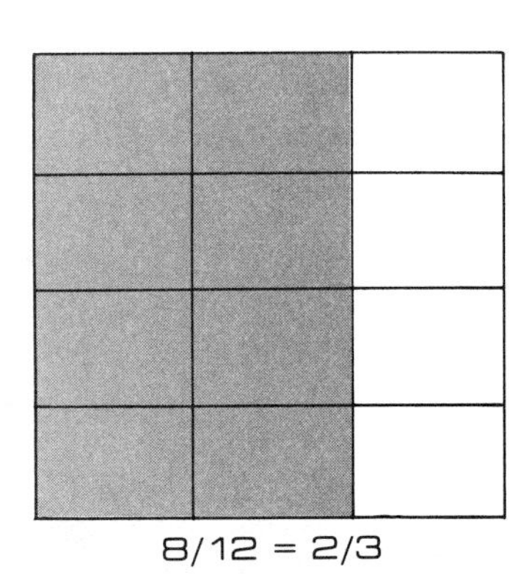

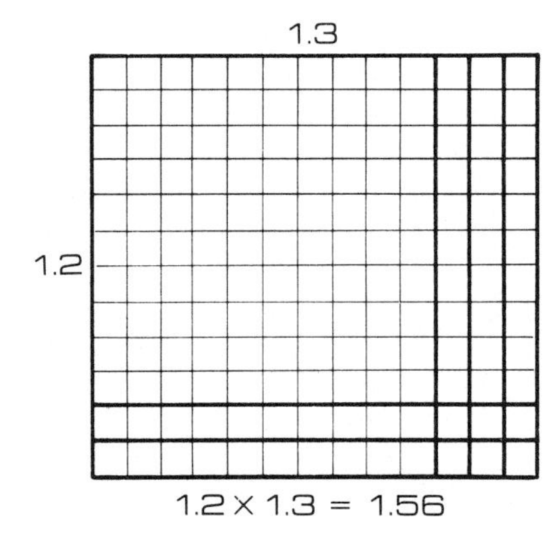

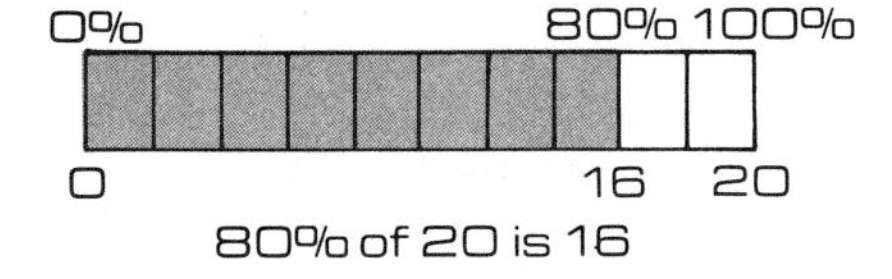

***Fig. 5.1.* Area models for fractions, decimals, and percents**

Later, students can extend area models to the study of algebra, probability, dimension analysis in measurement situations, and other more advanced subjects.

Concrete materials and representational models (pictures) should be used to continue the study of place value initiated in grades K–4. For example, students should see that the model in figure 5.2 represents 346 or 3.46, according to whether a small or a large square represents a unit.

Once students are familiar with a particular type of representation, it can be generalized and the equivalence of different representations for

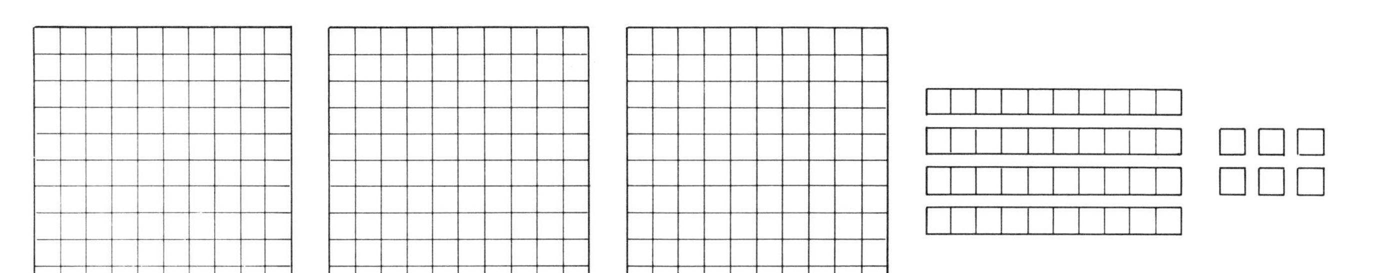

Fig. 5.2. Base-ten blocks

the same number can be established. This process should be approached conceptually before the computational techniques that convert one to another are developed.

In grades 5–8, number sense should be fostered through such questions as, How big is a million? or Could you carry a suitcase containing a million dollar bills? Operation sense should be expanded with such examples as, Is $2/3 \times 5/4$ more or less than $2/3$? More or less than $5/4$? Why is the product of a negative integer times a negative integer a positive integer?

Patterns that emerge when students examine terminating and repeating decimals are particularly appropriate for investigation in the middle grades. Questions about decimal expansions for fractions readily invite exploration: Which expansions are terminating, which are repeating, and which are nonrepeating? Which delay and then repeat? What is the relationship among expansions for families of fractions, such as $1/7$, $2/7$, $3/7$, . . ., $6/7$?

Just as K–4 students learn to represent single quantities with a number, middle school students should learn to represent comparison of quantities using ratios in various forms. Ratios should be introduced gradually through discussing the many situations in which they occur naturally. It takes little effort to relate these situations to students' interests: "If 245 of a company's 398 employees are women, how many of its 26 executives would you expect to be women?

Through these practical exercises, students should come to recognize that ratios are not directly measurable but that they contain two units and that the order of the items in the ratio pair in a proportion is critical. Thus, 23 persons per square mile is very different from 23 square miles per person. They need to understand the multiplicative nature of ratios to avoid such mistakes as "18 boys:15 girls :: 19 boys:16 girls" and to recognize nonexamples of ratios, for instance, doubling linear dimensions does not double area. They should begin to build other major ideas, including proportion, slope, and rational number. Ratios themselves can be extended to include more than two numbers, as in a recipe that has five ingredients. Spreadsheets provide an excellent means of working with ratios composed of more than two numbers, as illustrated in figure 5.3—students can see what happens to the amount of ingredients in a waffle recipe as the number of servings changes.

Joan's Waffles

	A	B	C	D	E
1	waffles	cups mix	cups milk	eggs	tb. oil
2	4	2	1.5	1	2
3	6	3	2.25	1.5	3
4					

Fig. 5.3. Spreadsheet for waffle recipe

Graphs can be used to show relationships involving numbers, including number line graphs and two-dimensional graphs, which can be expanded over the middle school years from whole number coordinates to rational numbers. For example, the following graph (fig. 5.4) comparing rainfall in two Canadian cities can generate discussions about the best time to visit the cities and other matters of interest.

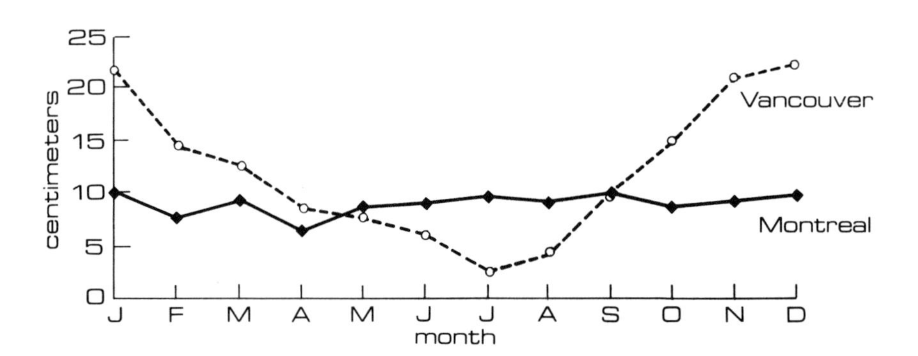

***Fig. 5.4.* Rainfall graph**

Over grades 5–8, students should build a sense of number and of numerical relationships that gives them the flexibility to deal with numbers in many forms.

STANDARD 6: NUMBER SYSTEMS AND NUMBER THEORY

In grades 5–8, the mathematics curriculum should include the study of number systems and number theory so that students can—

- ***understand and appreciate the need for numbers beyond the whole numbers;***
- ***develop and use order relations for whole numbers, fractions, decimals, integers, and rational numbers;***
- ***extend their understanding of whole number operations to fractions, decimals, integers, and rational numbers;***
- ***understand how the basic arithmetic operations are related to one another;***
- ***develop and apply number theory concepts (e.g., primes, factors, and multiples) in real-world and mathematical problem situations.***

Focus

The central theme of this standard is the underlying structure of mathematics, which bonds its many individual facets into a useful, interesting, and logical whole. Instruction in grades 5–8 typically devotes a great deal of time to helping students master a myriad of details but pays scant attention to how these individual facets fit together. It is the intent of this standard that students should come to understand and appreciate mathematics as a coherent body of knowledge rather than a vast, perhaps bewildering, collection of isolated facts and rules. Understanding this structure promotes students' efficiency in investigating the arithmetic of fractions, decimals, integers, and rationals through the unity of common ideas. It also offers insights into how the whole number system is extended to the rational number system and beyond. It improves problem-solving capability by providing a better perspective of arithmetic operations.

Instruction that facilitates students' understanding of the underlying structure of arithmetic should employ informal explorations and emphasize the reasons why various kinds of numbers occur, commonalities among various arithmetic processes, and relationships between number systems. Steen (1986, p. 6), past president of the Mathematical Association of America, articulated the spirit of this standard when he noted that "above all else, it [the mathematics curriculum] must not give the impression that mathematical and quantitative ideas are the product of authority or wizardry."

Number theory offers many rich opportunities for explorations that are interesting, enjoyable, and useful. These explorations have payoffs in problem solving, in understanding and developing other mathematical concepts, in illustrating the beauty of mathematics, and in understanding the human aspects of the historical development of number.

Discussion

As students reach grade 5, they begin to recognize—in both arithmetic and geometric settings—the need for numbers beyond whole numbers.

Arithmetically, the fraction $2/3$ becomes necessary as the only solution to the whole number problem $2 \div 3$, such as in the real-world situation in which two pizzas are divided among three people. The integer -1 becomes necessary so that the problem $2 - 3$ has a solution, such as when a player loses three points in a game when he or she has only two points. The need to measure more precisely than to the nearest inch gives rise to numbers like $3\frac{5}{8}$ inches. Such numbers as $\sqrt{2}$ and π are needed to describe the length of the diagonal of a square or the circumference of a circle. Encountering irrationals as nonexamples helps students appreciate rationals.

As students expand their mathematical horizons to include fractions, decimals, integers, and rational numbers, as well as the basic operations for each, they need to understand both the common ideas underlying these number systems and the differences among them. For example, to compare $2/3$ and $3/4$, students can use concrete materials to represent them as $8/12$ and $9/12$, respectively, and then to conclude that $8/12$ is less than $9/12$, since 8 is less than 9. Thus, they learn that comparing fractions is like comparing whole numbers once common denominators have been identified. Yet in comparing -2 and -5 on a number line or a thermometer, students see that -2 is greater than -5, so they learn that comparing negative integers is different from comparing whole numbers.

Students should extend their knowledge of whole number operations to other systems; for example, they need to see how $1.4 + 6.7$ is related to $14 + 67$ and how dividing -12 by -3 is related to dividing 12 by 3. Concrete materials and representational models, such as area models, should be used to provide a firm basis for understanding such ideas. The transition from whole numbers to fractions and decimals can be difficult for students. Although they may multiply the numerators and then the denominators, for example, they often do not understand why a similar procedure does not work in adding fractions. Concrete or representational models can help students clarify these anomalies.

Students should understand how analogies among structures can give a clearer picture of mathematics. For example, in contrasting the missing-addend interpretation of subtraction with the missing-factor interpretation of division, students learn that

$$12 - 3 = n \text{ means the same thing as } 3 + n = 12$$

and

$$12 \div 3 = n \text{ means the same thing as } 3 \times n = 12.$$

This suggests that subtraction can be considered in terms of addition and that division can be thought of in terms of multiplication, illustrating the analogy that division is to multiplication as subtraction is to addition.

Relationships among operations are developed over many grades. In K–4, multiplication can be viewed as repeated addition and division can be considered repeated subtraction. However, because the multiplication of fractions and decimals is not repeated addition and division is not always repeated subtraction, students should be exposed to other interpretations of multiplication and division as well. For example, as shown in figure 5.1, 1.2×1.3 can be represented by an area model. In the later grades, students learn that the subtraction of integers is the same as adding the opposite and the division of fractions is the same as multiplying by the reciprocal.

Another goal of this standard is to offer meaningful answers to such questions as, Why can't we divide by zero? Is there a smallest number? A largest number?

Challenging but accessible problems from number theory can be easily formulated and explored by students. For example, building rectangular arrays with a set of tiles can stimulate questions about divisibility and prime, composite, square, even, and odd numbers. (See fig. 6.1.)

Only one rectangle can be made with seven tiles, so 7 is prime.

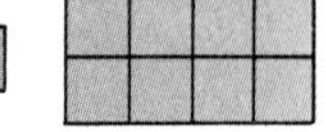

More than one rectangle can be made with eight tiles, so 8 is composite.

***Fig. 6.1.* Tile explorations**

This activity and others can be extended to investigate other interesting topics, such as abundant, deficient, or perfect numbers; triangular and square numbers; cubes; palindromes; factorials; and Fibonacci numbers. The development of various procedures for finding the greatest common factor of two numbers can foreshadow important topics in the 9–12 curriculum, as students compare the advantages, disadvantages, and efficiency of various algorithms. String art and explorations with star polygons can relate number theory to geometry.

Another example from number theory involves making connections between the prime structure of a number and the number of its factors:

Find five examples of numbers that have exactly three factors. Repeat for four factors, then five factors. What can you say about the numbers in each of your lists?

Students might give 4, 9, 25, 49, and 121 as examples of numbers with exactly three factors. Each of these numbers is the square of a prime.

Without an understanding of number systems and number theory, mathematics is a mysterious collection of facts. With such an understanding, mathematics is seen as a beautiful, cohesive whole.

STANDARD 7: COMPUTATION AND ESTIMATION

In grades 5–8, the mathematics curriculum should develop the concepts underlying computation and estimation in various contexts so that students can—

- ***compute with whole numbers, fractions, decimals, integers, and rational numbers;***
- ***develop, analyze, and explain procedures for computation and techniques for estimation;***
- ***develop, analyze, and explain methods for solving proportions;***
- ***select and use an appropriate method for computing from among mental arithmetic, paper-and-pencil, calculator, and computer methods;***
- ***use computation, estimation, and proportions to solve problems;***
- ***use estimation to check the reasonableness of results.***

Focus

Although computation is vital in this information age, technology has drastically changed the methods by which we compute. Whereas inexpensive calculators execute routine computations accurately and quickly and computers execute more complex computations with ease, many current mathematics programs focus on traditional paper-and-pencil algorithms. This standard prepares students to select and use appropriate mental, paper-and-pencil, calculator, and computer methods.

It is no longer necessary or useful to devote large portions of instructional time to performing routine computations by hand. Other mathematical experiences for middle school students deserve far more emphasis. The facility in computation that calculators and computers offer should imbue the curriculum with an expanded potential for interesting problem-solving experiences. Students need more experience in developing procedures and evaluating their work and in interpreting the results of computations done by machines.

It is beneficial to view computation in relation to other topics in the mathematics curriculum. Computation, estimation, or methods for solving proportions should not be considered or taught as ends in themselves. In grades 5–8, computation and estimation should be integrated with the study of the concepts underlying fractions, decimals, integers, and rational numbers, as well as with the continuing study of whole number concepts. Similarly, methods for solving proportions should be viewed as one aspect of students' growing but informal understanding of ratio and proportions. These computational procedures should be developed in context so that the learner perceives them as tools for solving problems.

Discussion

This standard extends the K–4 standard on computation with whole numbers to include new number systems. Broadly, its premise is that computation should support meaningful experiences in geometry, proba-

♦ ♦ ♦ ♦ ♦ ♦ ♦ ♦

bility, measurement, and other areas of mathematics. Estimation should be used to solve problems for which exact answers are inappropriate and to check computation results. This more general notion of computation is an important part of mathematical communication. In the middle grades, computation requires that students be able to select a symbol system appropriate to a particular context and represent an idea, a solution to a problem, or a particular situation with a symbolic procedure.

The greatest revisions to be made in the teaching of computation include the following:

- Fostering a solid understanding of, and proficiency with, simple calculations
- Abandoning the teaching of tedious calculations using paper-and-pencil algorithms in favor of exploring more mathematics
- Fostering the use of a wide variety of computation and estimation techniques—ranging from quick mental calculation to those using computers—suited to different mathematical settings
- Developing the skills necessary to use appropriate technology and then translating computed results to the problem setting
- Providing students with ways to check the reasonableness of computations (number and algorithmic sense, estimation skills)

The ability to compute 0.17×45 correctly is not as interesting for its own sake as it is for estimating the number of times a certain result will occur on a spinner in a game or in determining a discount when buying a new tennis racquet. Students should know when it is appropriate to multiply 0.17×45 in problem-solving situations and how to multiply 0.17×45 with a calculator.

Despite these fundamental revisions, certain aspects of computation continue to be important. A knowledge of basic facts and procedures is critical in mental arithmetic and estimation. Knowing that $8 \times 7 = 56$ is a basis for finding 8×700 mentally, multiplying $(+8) \times (-7)$, estimating 824×689, and estimating 8.24×6.89.

Valuable class time should not be devoted to developing students' proficiency in calculating 824×689 or 8.24×6.89 with paper and pencil, since these exercises can be done more readily with a calculator.

As they begin to understand the meaning of operations and develop a concrete basis for validating symbolic processes and situations, students should design their own algorithms and discuss, compare, and evaluate them with their peers and teacher. Students should analyze the way the various algorithms work and how they relate to the meaning of the operation and to the numbers involved.

Similar experiences can stimulate mental arithmetic, a particularly good topic for enhancing students' understanding of numbers and their sense of power in mathematics. In the following example, a teacher asks students to add in their heads the number of cans collected this week (157) to the number already collected (1950) for sale to a recycling depot.

Tony: **2007.**

Julie: **2107.**

Ms. Clark: **Tony, how did you do it?**

♦ ♦ ♦ ♦ ♦ ♦ ♦ ♦

Tony: **Well, I added 50 to 1950 and got 2000. Then I put on . . . Oh! Oh! I forgot the hundred.**

Ms. Clark: **Interesting. Julie?**

Julie: **I got 21 hundreds: 1 + 19 and 1 more from 50 + 50. Then I added 7.**

Ms. Clark: **O.K., Julie, so we'll have 2107 cans. How many boxes will we need to carry them out to the truck?**

This example illustrates many points. First, computation arises from the need to do or know, not simply for its own sake. Second, students can and do make many interesting algorithms. Third, erroneous results offer opportunities for learning when students can describe what they are doing and can check their answers. The teacher can follow this exercise with a discussion of the role of the numeration system in the students' mental computations and extend the lesson to computation itself: "Would the same methods work for computing the sum of three items in a store: $19.50, $1.57, and $22.00?"

The mastery of a small number of basic facts with common fractions (e.g., $1/4 + 1/4 = 1/2$; $3/4 + 1/2 = 1\,1/4$; and $1/2 \times 1/2 = 1/4$) and with decimals (e.g., $0.1 + 0.1 = 0.2$ and $0.1 \times 0.1 = 0.01$) contributes to students' readiness to learn estimation and for concept development and problem solving. This proficiency in the addition, subtraction, and multiplication of fractions and mixed numbers should be limited to those with simple denominators that can be visualized concretely or pictorially and are apt to occur in real-world settings; such computation promotes conceptual understanding of the operations. This is not to suggest, however, that valuable instruction time should be devoted to exercises like $17/24 + 5/18$ or $5\,3/4 \times 4\,1/4$, which are much harder to visualize and unlikely to occur in real-life situations. Division of fractions should be approached conceptually. An understanding of what happens when one divides by a fractional number (less than or greater than 1) is essential.

Similarly, students should learn to compute decimal products like 0.3×0.6, especially as a means of locating the decimal point. Although such problems train students to estimate more difficult computations, valuable instructional time should not be devoted to calculating products such as 0.31×0.588 with paper and pencil.

Instruction should stress informal but effective methods for solving proportions, including ways to identify integer multiples in ratios and processes in which changing one of the ratios to an equivalent unit ratio is an intermediate step. Cross-multiplication methods should be deferred until students understand these methods in algebraic terms.

Basic facts, processes, and translations within and among common and decimal fractions, percents, proportions, and integers are important to students' understanding of computation. Performing two-digit computations with whole numbers or decimals aids students in understanding connections between computation and numeration. Even though students can explore paper-and-pencil computations with numbers of any size and with various systems, they should not be expected to become proficient with paper-and-pencil computations with several digits. A curriculum that incorporated this standard would not include paper-and-pencil practice for proficiency with tedious computations, such as those with three-digit multipliers or divisors, or with operations on fractions beyond the extent suggested previously. Students should possess adequate mental arithmetic skills so that they are not dependent on calculators to do simple computations and are able to detect unreasonable an-

◆ ◆ ◆ ◆ ◆ ◆ ◆ ◆

swers when using calculators to solve harder computations. But this standard concentrates to a far greater degree on teaching students to use computations in context, to frame and execute computations using different methods, and to estimate.

Appropriate activities can build in students the kind of computational sense, capability, and confidence intended by this standard. Here is one example:

A set of cards is prepared, each one bearing the price of an object and a particular discount in percentages (e.g., $10.95, 15%). Each of the two players has a calculator. One player turns over a card to reveal a price and a discount. Then both players estimate the final, discounted price. They use the calculators to find the discounted price, and the player who comes closest to the actual discounted price earns one point. A game played to ten points takes ten minutes or less.

Choosing an appropriate method for performing a computation is more important than it has been in the past. For example, consumers might use mental arithmetic to estimate the amount of a tip in a restaurant, use a calculator to fill out a tax form, or use a computer to determine how long it will take to pay off a credit-card balance. Students need to learn how to do each method of computation and to choose which one to use in a given situation. Students should translate results from various computational devices into solutions that fit particular problems and settings. For example, consider the following situation:

A tray can hold 12 salads. How many trays are required for 244 persons?

If a student's calculator shows the answer 20.333333, he or she must be able to interpret this as 20 full trays plus 1 partially filled tray (or 21 trays).

Similar interpretations are required with computer software. If students use a spreadsheet to represent pay rates for newspaper carriers, they need to learn how to relate the information on the screen or printout to decisions they are trying to make.

Estimation is a powerful mathematical idea to be used both in solving problems and in checking the reasonableness of results. When a student wants to know about how long it will take to earn enough baby-sitting money to buy a new bicycle, he or she can estimate the answer. The use of basic facts as well as various techniques for estimation (e.g., rounding and comparison) are used in the following examples to check the reasonableness of results:

Sarah's calculator showed the answer 676.8 after she multiplied 9.4 × 7.2 (she forgot to enter the decimal point in 7.2). Her estimate suggested the answer should be about 9 × 7, or about 63.

Hector got $\frac{4}{6}$ when he added $\frac{3}{4} + \frac{1}{2}$ (he confused the rules for adding fractions with those for multiplying fractions and added the numerators and then the denominators). His estimate suggested the answer should be greater than $\frac{1}{2} + \frac{1}{2}$, or larger than 1.

It is the intent of this standard that computation be viewed not as a goal in itself but as a multifaceted tool for knowing and doing.

STANDARD 8: PATTERNS AND FUNCTIONS

In grades 5–8, the mathematics curriculum should include explorations of patterns and functions so that students can—

- ***describe, extend, analyze, and create a wide variety of patterns;***
- ***describe and represent relationships with tables, graphs, and rules;***
- ***analyze functional relationships to explain how a change in one quantity results in a change in another;***
- ***use patterns and functions to represent and solve problems.***

Focus

One of the central themes of mathematics is the study of patterns and functions. This study requires students to recognize, describe, and generalize patterns and build mathematical models to predict the behavior of real-world phenomena that exhibit the observed pattern. The widespread occurrence of regular and chaotic pattern behavior makes the study of patterns and functions important. Exploring patterns helps students develop mathematical power and instills in them an appreciation for the beauty of mathematics.

The study of patterns in grades 5–8 builds on students' experiences in K–4 but shifts emphasis to an exploration of functions. However, work with patterns continues to be informal and relatively unburdened by symbolism. Students have opportunities to generalize and describe patterns and functions in many ways and to explore the relationships among them. When students make graphs, data tables, expressions, equations, or verbal descriptions to represent a single relationship, they discover that different representations yield different interpretations of a situation. In informal ways, students develop an understanding that functions are composed of variables that have a dynamic relationship: Changes in one variable result in change in another. The identification of the special characteristics of a relationship, such as minimum or maximum values or points at which the value of one of the variables is 0 (x- and y-intercepts), lays the foundation for a more formal study of functions in grades 9–12.

The theme of patterns and functions is woven throughout the 5–8 standards. It begins in K–4, is extended and made more central in 5–8, and reaches maturity with a natural extension to symbolic representation and supporting concepts, such as domain and range, in grades 9–12. Examples appropriate for grades 5–8 are incorporated into other standards for this age group.

Discussion

During the middle years, the study of patterns and functions should focus on the analysis, representation, and generalization of functional relationships. These topics should first be explored as informal investigations.

Students should be encouraged to observe and describe all sorts of patterns in the world around them: plowed fields, haystacks, architecture, paintings, leaves on trees, spirals on pineapples, and so on. As the students mature, instructional efforts can move toward building a firm grasp of the interplay among tables of data, graphs, and algebraic expressions as ways of describing functions and solving problems.

Many problems challenge students to find clever ways to find a solution by counting. Looking for patterns in simple situations can lead to a method of counting generalizable to other situations. For example:

Good news travels fast. Iris saved enough money from her paper route to buy a new bicycle. She immediately told two friends, who, ten minutes later, each repeated the news to two other friends. Ten more minutes later, these friends each told two others. If the news continues to spread in this fashion, how many people will know about Iris's new bicycle after eighty minutes?

Time	0	10	20	30	40	5
People told	2	4	8	16	32	
Total	2	6	14	30		

***Fig. 8.1.* Good news table**

Students can approach this problem in different ways. One way is to organize specific cases into a table (fig. 8.1).

Each row of the table represents a different function. The "Time" row is a linear function involving multiples of 10. The "People told" row is basically an exponential function involving powers of 2. The pattern in the "Total" row can be seen in different ways. One is to observe that a new entry can be found by adding the current entry in "Total" to the next entry in "People told" as illustrated by the arrows in the table.

Students can use their calculators to find the entries for specific times. More mature students can be challenged to find an algebraic rule that will describe the total for any time (t). Students can also be challenged to develop another representation of the data, such as a graph.

The following illustrates a problem situation involving patterns that can be tackled at many levels:

Investigate what happens when different-sized cubes are constructed from unit cubes, the surface area is painted, and the large cube is then disassembled into its original unit cubes. How many of the $1 \times 1 \times 1$ cubes are painted on three faces, two faces, one face, and no faces?

Students can approach this problem by looking at particular instances, organizing the data in a table, looking for patterns, generalizing the patterns, and graphing the four relationships. See figure 8.2.

		# of unit cubes with paint on			
Dimensions	**# of 1 × 1 × 1 cubes needed**	**3 faces**	**2 faces**	**1 face**	**0 faces**
2 × 2 × 2	8	8	0	0	0
3 × 3 × 3	27	8	12	6	1
4 × 4 × 4	64	8	24	24	8
5 × 5 × 5	125	8	36	54	27
.	.	.	.	.	.
.	.	.	.	.	.
.	.	.	.	.	.
$n \times n \times n$	n^3	8	$12(n-2)$	$6(n-2)^2$	$(n-2)^3$

***Fig. 8.2.* Cube painting**

An interesting feature of this problem is that it involves linear, constant, quadratic, and cubic functions. For example, asking whether 3174 will appear in the "painted on one face" column allows students to apply the number-theory concepts of divisibility and square numbers informally to solve quadratic equations in a situation that gives meaning to the concepts and solution.

Much of mathematics in grades 5–8 can be viewed as an exploration of patterns and regularity. Pi is best understood by students when they investigate the ratio between circumference and diameter by measuring round objects and looking for regularities in the data or the graph of circumferences compared to diameters. The number of diagonals of any polygon becomes predictable when an exploration of patterns reveals the underlying function. Integers and operations on integers become a natural extension of whole numbers when viewed in terms of patterns:

$3 \times 2 = 6$	$3 \times -2 = -6$
$3 \times 1 = 3$	$2 \times -2 = -4$
$3 \times 0 = 0$	$1 \times -2 = -2$
$3 \times -1 = -3$	$0 \times -2 = 0$
$3 \times -2 = -6$	$-1 \times -2 = 2$

The following situation encourages students to reason from patterns to solve a real-world problem.

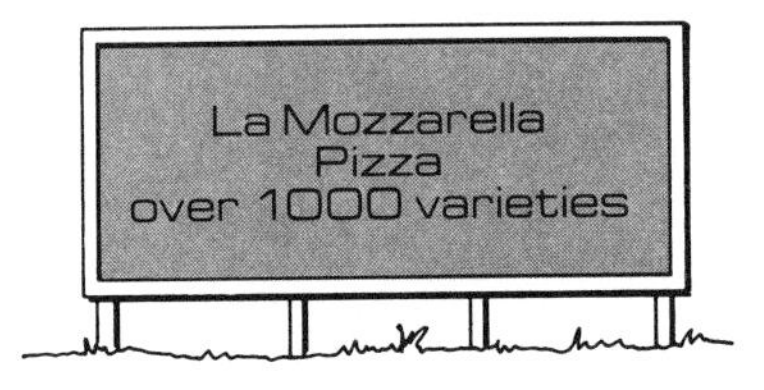

***Fig. 8.3.* Pizza patterns**

Pat saw a billboard along the roadside (fig. 8.3). If a plain pizza has just cheese and tomato, how many toppings does La Mozzarella's have?

Students can explore this problem by taking different numbers of toppings and listing each possible pizza. This can lead to the development of Pascal's triangle and an exploration of the many other situations that relate to it.

In the following problem, students need to generate, organize, and analyze data; look for patterns; and then use the observed patterns to generalize.

In a village are 3 streets. All the streets are straight. Each crossroad has one lamppost. How many lampposts are needed? How many are needed for a village with 20 streets? Generalize to any number of streets.

This problem is not well formed. Questions about how the streets are laid out must be answered before students can pursue their own solutions. As indicated by all the preceding problem situations, the dynamic relationship between variables in a pattern or function can be viewed physically and in tables of data. Here are some ways to explore change represented in graphs:

Mary and her brother John leave home together to walk to school. Mary thinks they are going to be late, so she starts out running, then tires and walks the rest of the way. John starts out walking and starts to run as he nears the school building. They arrive at the same time. The graphs in figure 8.4 show the distance from their home on the vertical axis and time on the horizontal axis. Which graph best represents Mary's trip? John's trip?

Patterns abound in our world. The mathematics curriculum should help sensitize students to the patterns they meet every day and to the math-

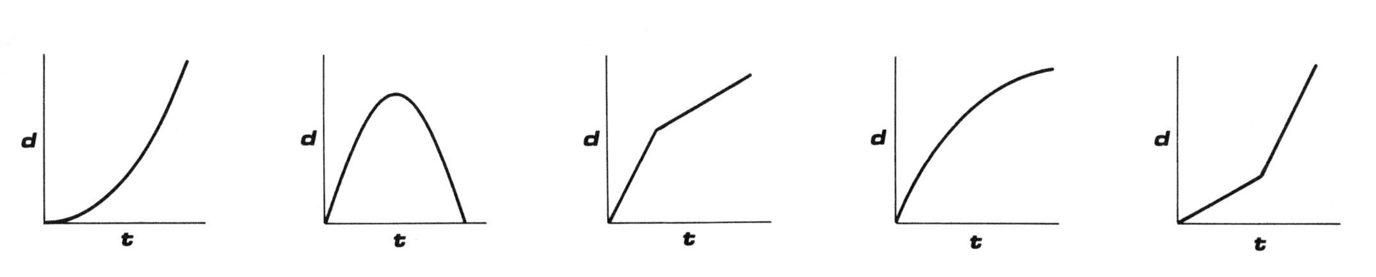

Fig. 8.4. **Home-to-school graph: distance versus time**

ematical descriptions or models of these patterns and relationships. A mathematical model's ability to predict is a powerful, fundamental mathematical concept that deserves continuing emphasis in the curriculum.

STANDARD 9: ALGEBRA

In grades 5–8, the mathematics curriculum should include explorations of algebraic concepts and processes so that students can—

- ***understand the concepts of variable, expression, and equation;***
- ***represent situations and number patterns with tables, graphs, verbal rules, and equations and explore the interrelationships of these representations;***
- ***analyze tables and graphs to identify properties and relationships;***
- ***develop confidence in solving linear equations using concrete, informal, and formal methods;***
- ***investigate inequalities and nonlinear equations informally;***
- ***apply algebraic methods to solve a variety of real-world and mathematical problems.***

Focus

The middle school mathematics curriculum is, in many ways, a bridge between the concrete elementary school curriculum and the more formal mathematics curriculum of the high school. One critical transition is that between arithmetic and algebra. It is thus essential that in grades 5–8, students explore algebraic concepts in an informal way to build a foundation for the subsequent formal study of algebra. Such informal explorations should emphasize physical models, data, graphs, and other mathematical representations rather than facility with formal algebraic manipulation. Students should be taught to generalize number patterns to model, represent, or describe observed physical patterns, regularities, and problems. These informal explorations of algebraic concepts should help students to gain confidence in their ability to abstract relationships from contextual information and use a variety of representations to describe those relationships.

Activities in grades 5–8 should build on students' K–4 experiences with patterns. They should continue to emphasize concrete situations that allow students to investigate patterns in number sequences, make predictions, and formulate verbal rules to describe patterns. Learning to recognize patterns and regularities in mathematics and make generalizations about them requires practice and experience. Expanding the amount of time that students have to make this transition to more abstract ways of thinking increases their chances of success. By integrating informal algebraic experiences throughout the K–8 curriculum, students will develop confidence in using algebra to represent and solve problems. In addition, by the end of the eighth grade, students should be able to solve linear equations by formal methods and some nonlinear equations by informal means.

Discussion

Understanding the concept of *variable* is crucial to the study of algebra; a major problem in students' efforts to understand and do algebra results from their narrow interpretation of the term. Many students assign a numerical value to a letter from the start; others ignore the letter, and still others treat the letter as shorthand for an object (*b* means *boy* rather than *number of boys*). Students need to be able to use variables in many ways. Two particularly important ways in grades 5–8 are using a variable as a placeholder for a specific unknown, as in $n + 5 = 12$, and as a representative of a range of values, as in $3t + 6$. Students who work with computers are likely to encounter the replacement use of variables. Students need to tell from the context how a variable is being used.

Giving students opportunities to explore interesting problems, applications, and situations does not guarantee that they will make the appropriate connections; it is inevitable that some students might lose sight of the important mathematical ideas that underlie any activity. They need to be encouraged and helped to reflect on their explorations and summarize concepts, relationships, processes, and facts that have emerged from their discussions.

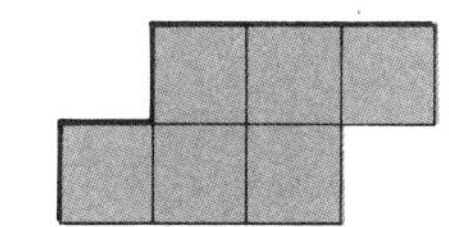

Fig. 9.1. **Tile shapes**

The following example illustrates how students can develop a sophisticated understanding of how algebra can be used to model situations and how the algebraic model is related to other models or representations:

Working with square tiles, students can explore the question, "Can you add tiles to this figure [see fig. 9.1] to make a new figure with a perimeter of 18 units?" (Tiles must touch each other along an entire edge.)

Fig. 9.2. **Rectangles**

Students can discover many interesting facts and relationships in exploring this problem. They can discover that adding a tile to fill in a corner where it will touch other tiles along two edges does not change the perimeter at all; that adding a tile that touches another tile along one edge changes the perimeter by exactly two units; and that adding a tile so that it touches three edges actually reduces the perimeter. The students can write algebraic expressions to summarize their discoveries, for example, $p + 2$ or $p - 2$ for adding tiles that touch one or three edges, respectively.

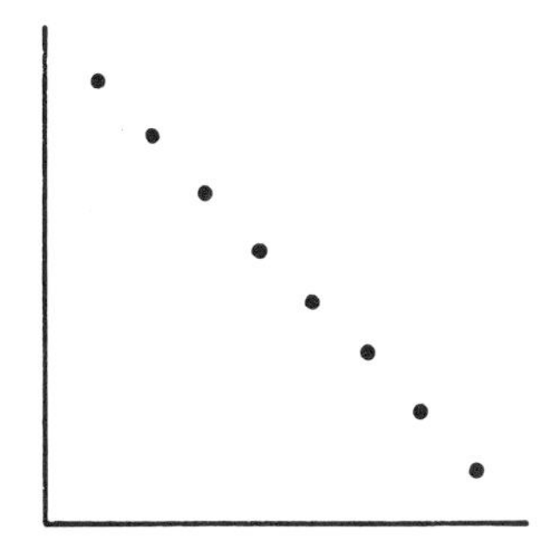

Fig. 9.3. **Width versus length**

Once the class has found different ways to add tiles to make a new figure with a perimeter of 18, students can explore other problems, such as determining the fewest number of tiles that can be added. Students discover that at least three tiles must be added before the perimeter will total 18 units. A question about the greatest number of tiles that can be added to reach a perimeter of 18 units leads to an interesting discovery: A rectangle that is 4 tiles by 5 tiles uses the most tiles. This raises the question, What other rectangles have perimeters of 18 units? Collecting and organizing the class data yields a table of values for further investigation.

Fifth-grade students might cut physical representations of the rectangles from grid paper. These can be stacked on a second sheet of grid paper to produce a physical graph of the relationship between length and width (see fig. 9.2).

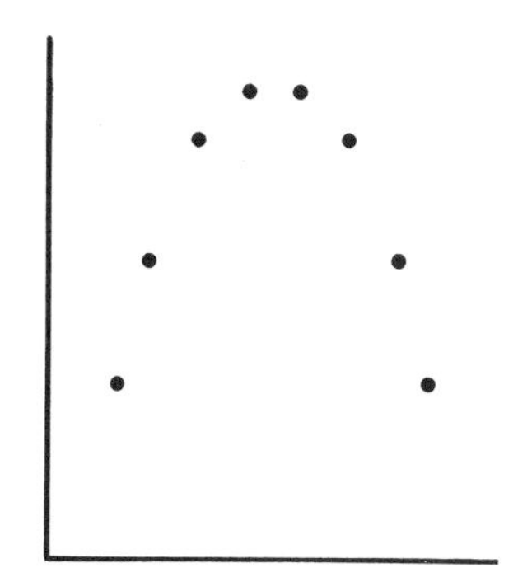

Fig. 9.4. **Area versus length**

In later grades, students can locate points representing the length versus the width of each rectangle (see fig. 9.3). Teachers can ask such questions as, Does it make sense to connect the points in the graph? How would you interpret the points where the line intersects the horizontal axis? If you can use fractional measures for the sides of the rectangle with a perimeter of 18 units, what width would you expect to find for a length between 4 and 5 units? What would you say about its area? Students can make a graph to picture the relationship between the length and the area (see fig. 9.4).

Other questions to explore include, What happens to the curve at the top of the graph? Can you find the rectangle with the maximum area for a fixed perimeter of 18 units? These questions can help students discover that the rectangle with the maximum area for a fixed perimeter is a square. Here the students have explored in a concrete way an idea that is fundamental to calculus: finding maximum and minimum values.

From these concrete, numerical, and graphic models emerges an algebraic model of the relationships, $2L + 2W = 18$ or $W = 9 - L$ and the rule for area of a rectangle, $A = LW$. Thus $A = L(9 - L)$ or $A = 9L - L^2$. Connecting $A = 9L - L^2$ to the graph of the data points for different values of L is a very important result of this exploration and discussion.

Relating models to one another builds a better understanding of each. Every step in these examples helps students develop an understanding of symbolic representation: exploring a concrete situation to determine patterns, constructing a table of data, looking for ways to generalize the situation described by the table, asking questions about how the variables are related, making a graphical representation, and looking for maximum and minimum points, or points where the graph intersects the axes. Different problem situations call for different approaches. Appropriate situations for exploration can arise from mathematical or real-world contexts. In either case, problem solving of this sort enables students to develop confidence in their ability to use algebraic ideas.

Many informal ways of studying algebraic expressions and solving linear equations can arise in such contexts. In the number trick illustrated in Standard 2 (see fig. 2.1), tiles represent the "any number" notion of variable and sets of beans represent known numbers. Variables can be introduced as names for the tiles. Tile and bean combinations can be transformed: Pick a tile, add 5, double the result. Students should act out these directions and be able to describe them in words and symbols: $n + 5$; $2 \times n + 10$. The association of language, materials, and actions builds intuitions of algebra. Such algebraic and "tile" expressions can be assigned different values by asking each student in a group to give a value for "any number" (the tile) and then evaluate the expression. This can lead to guess-and-test equation solving with students challenged to find the value of one tile if two tiles plus 10 is 20.

A computer also can be very effective in exploring relationships between two expressions in a real-world context. In a problem about newspaper deliveries, P represents the number of papers delivered. One paper pays 50 cents a day plus 8 cents for each paper delivered; another pays 10 cents for each paper delivered. The computer can generate a set of values for P and related values for $8 \times P + 50$ and $10 \times P$, leading to the discussion and solving of $8 \times P + 50 = 10 \times P$, or $8 \times P + 50 < 10 \times P$ in informal ways.

Formal equation-solving methods can be developed from, and supported by, informal methods. These informal methods, which may include actions on concrete materials that are paralleled by symbolic actions, can lead to more formal procedures. If students develop formal procedures from informal methods grounded in real-world contexts, they can validate their own formal thinking and develop a basis for extending these algebraic ideas.

STANDARD 10: STATISTICS

In grades 5–8, the mathematics curriculum should include exploration of statistics in real-world situations so that students can—

- ***systematically collect, organize, and describe data;***
- ***construct, read, and interpret tables, charts, and graphs;***
- ***make inferences and convincing arguments that are based on data analysis;***
- ***evaluate arguments that are based on data analysis;***
- ***develop an appreciation for statistical methods as powerful means for decision making.***

Focus

In this age of information and technology, an ever-increasing need exists to understand how information is processed and translated into usable knowledge. Because of society's expanding use of data for prediction and decision making, it is important that students develop an understanding of the concepts and processes used in analyzing data. A knowledge of statistics is necessary if students are to become intelligent consumers who can make critical and informed decisions.

In grades K–4, students begin to explore basic ideas of statistics by gathering data appropriate to their grade level, organizing them in charts or graphs, and reading information from displays of data. These concepts should be expanded in the middle grades. Students in grades 5–8 have a keen interest in trends in music, movies, fashion, and sports. An investigation of how such trends are developed and communicated is an excellent motivator for the study of statistics. Students need to be actively involved in each of the steps that comprise statistics, from gathering information to communicating results.

Identifying the range or average of a set of data, constructing simple graphs, and reading data points as answers to specific questions are important activities, but they reflect only a very narrow aspect of statistics. Instead, instruction in statistics should focus on the active involvement of students in the entire process: formulating key questions; collecting and organizing data; representing the data using graphs, tables, frequency distributions, and summary statistics; analyzing the data; making conjectures; and communicating information in a convincing way. Students' understanding of statistics is also enhanced by evaluating others' arguments. This exercise is of particular importance to all students, since advertising, forecasting, and public policy are frequently based on data analysis.

Discussion

Middle school students' curiosity about themselves, their peers, and their surroundings can motivate them to study statistics. The data to be gathered, organized, and studied should be interesting and relevant; students' interest in themselves and their peers, for example, can motivate them to investigate the "average" student in the class or school. First, students can formulate questions to determine the characteristics of an "average" student—age, height, eye color, favorite music or TV show, number of people in family, pets at home, and so on. Although numerous categories are possible, some discussion will help students to develop a survey instrument to obtain appropriate data. Sampling procedures are a critical issue in data collection. Which students should be surveyed to determine Mr. or Ms. Average? Must every student be questioned? If not, how can randomness in the sampling be assured and how many samples are needed to accumulate enough data to describe the average student?

Random samples, bias in sampling procedures, and limited samples all are important considerations. For instance, would collecting data from the men's and women's basketball teams provide needed information to determine the average height of a college student? Will a larger sample reveal a more accurate picture of the percentage of students with brown hair? The graph in figure 10.1 illustrates the results of increasing the sample size.

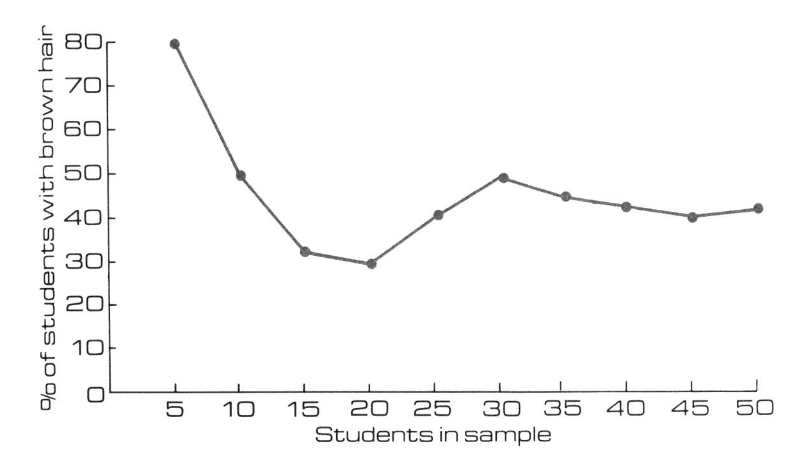

Fig. 10.1. Graph of brown-haired students

Data can be presented in many forms: charts, tables (fig. 10.2), plots (e.g., stem-and-leaf, box-and-whiskers, and scatter), and graphs (e.g., bar, circle, or line). Each form has a different impact on the picture of the information being presented, and each conveys a different perspective. The choice of form depends on the questions that are to be answered. Using the same data, students can develop graphs with different scales to show how the change of scale can dramatically alter the visual message that is communicated.

Number with brown hair	4	5	5	6	10	15	16	17	18	21
Students sampled	5	10	15	20	25	30	35	40	45	50

Fig. 10.2. Data table

Computer software can greatly enhance the organization and representation of data. Data-base programs offer a means for students to structure, record, and investigate information; to sort it quickly by various categories; and to organize it in a variety of ways. Other programs can be used to construct plots and graphs to display data. Scale changes can be made to compare different views of the same information. These technological tools free students to spend more time exploring the essence of statistics: analyzing data from many viewpoints, drawing inferences, and constructing and evaluating arguments.

A particular point that should be raised with students is how *average* relates to numerical and nonnumerical data. Although there are several measures of central tendency, students are generally exposed only to the mean or median, yet the mode might be the best "average" for a set of nonnumerical data.

Students should also explore the concepts of *center* and *dispersion of data.* The following activity includes all the important elements of this standard and illustrates the use of box-and-whisker plots as an effective means of describing data and showing variation.

A class is divided into two large groups and then subdivided into pairs. One student in each pair estimates when one minute has passed, and the other watches the clock and records the actual time. All the students in one group concentrate on the timing task, while half the students in the second group exert constant efforts to distract their partners. The box plots show that the median times for the two groups are about the same but the times for the distracted group have greater variation. Note that in the distracted group, one data point is far enough removed from the others to be an outlier. See figure 10.3.

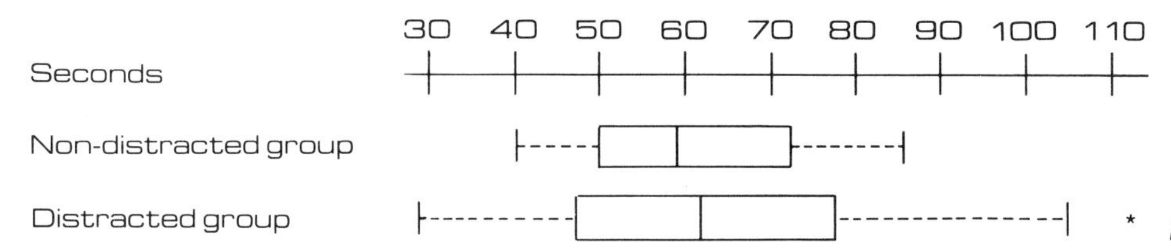

***Fig. 10.3.* Time estimates**

Sports statistics and other real data are settings in which students can generate new data and investigate a variety of conjectures. The table in figure 10.4 contains some information from an NBA championship game between Los Angeles and Boston.

Player	Minutes Played	Field Goals/ Attempts	Rebounds	Assists	Points
Worthy	37	8/19	8	5	20
Johnson	34	8/14	1	12	19
Bird	31	8/14	6	9	19
McHale	32	10/16	9	0	26

***Fig. 10.4.* NBA championship game statistics**

Using the table, students can generate such new information as points/minute, rebounds/minute, points/field goals attempted. Who is the best percentage shooter? From another source, they can find the height of each player and determine rebounds/inch of height or points/inch of height. A problem like this is ideally suited to the curious nature of middle school students and opens up a world of questions and investigations to them.

Formulating key questions, interpreting graphs and charts, and solving problems are important goals in the study of statistics. Statistics can help answer questions that do not lend themselves to direct measurement. Once data are collected and organized, such questions as the following can guide students in interpreting the data:

- What appears most often in the data?
- What trends appear in the data?

- What is the significance of outliers?
- What interpretations can we draw from these data, and can we use our interpretations to make predictions?
- What difficulties might we encounter when extending the interpretations or predictions to other related problems?
- What additional data can we collect to verify or disprove the ideas developed from these data?

All media are full of graphical representations of data and different kinds of statistical claims that can be used to stimulate discussion of the message conveyed and the arguments presented in the data.

STANDARD 11: PROBABILITY

In grades 5–8, the mathematics curriculum should include explorations of probability in real-world situations so that students can—

- ***model situations by devising and carrying out experiments or simulations to determine probabilities;***
- ***model situations by constructing a sample space to determine probabilities;***
- ***appreciate the power of using a probability model by comparing experimental results with mathematical expectations;***
- ***make predictions that are based on experimental or theoretical probabilities;***
- ***develop an appreciation for the pervasive use of probability in the real world.***

Focus

> Probability theory is the underpinning of the modern world. Current research in both the physical and social sciences cannot be understood without it. Today's politics, tomorrow's weather report and next week's satellites all depend on it. (Huff and Geise 1959)

An understanding of probability and the related area of statistics is essential to being an informed citizen. Often we read statements such as, "There is a 20 percent chance of rain or snow today." "The odds are three to two that the Cats will win the championship." "The probability of winning the grand prize in the state lottery is 1 in 7 240 000." Students in the middle grades have an intense interest in the notions of fairness and the chances of winning games. The study of probability develops concepts and methods for investigating such situations. These methods allow students to make predictions when uncertainty exists and to make sense of claims that they see and hear.

The study of probability in grades 5–8 should not focus on developing formulas or computing the likelihood of events pictured in texts. Students should actively explore situations by experimenting and simulating probability models. Such investigations should embody a variety of realistic problems, from questions about sports events to whether it will rain on the day of the school picnic. Students should talk about their ideas and use the results of their experiments to model situations or predict events. Probability is rich in interesting problems that can fascinate students and provide settings for developing or applying such concepts as ratios, fractions, percents, and decimals.

Discussion

Probability, the measure of the likelihood of an event, can be determined theoretically or experimentally. Students in the middle grades must actively participate in experiments with probability so that they develop an understanding of the relationship between the numerical expression of a probability and the events that give rise to these numbers (e.g., $2/5$ as it relates to the probability of choosing a red marble from a hat). Students must not only understand the relationship between the numerical expression and the probability of the events but realize that the measure of certainty or uncertainty varies as more data are collected.

Students have many misconceptions and poor intuitions about probabilistic situations. In order to bring these ill-formed notions to the conscious level so that they can be confronted, students should be asked to guess what will happen next or what the result of the experiment will show. An unexpected result has a much greater potential to cause students to rethink their basic assumptions if they have articulated their ideas before their experiment or analysis of the situation.

To see how the predictions we hear and see every day are based on probability, students must use their knowledge of probability to solve problems. In modeling problems, conducting simulations, and collecting, graphing, and studying data, students will come to understand how predictions can be based on data. Mathematically derived probabilities can be determined by building a table or tree diagram, creating an area model, making a list, or using simple counting procedures. Students develop an appreciation of the power of simulation and experimentation by comparing experimental results to the mathematically derived probabilities. For example, students can conduct experiments with two dice to determine the experimental probability of rolling a 4, construct a table to establish the theoretical probability, and then compare the two results. The following activity includes all these concepts and an analysis of fairness.

Arrange students in pairs and give each pair three chips: one chip with an "A" on one side and a "B" on the other; a second with "A" on one side and "C" on the other; and the third with "B" on one side and "C" on the other. One student tosses all three chips simultaneously onto the desk. Player 1 wins if any two chips match; Player 2 wins if all three chips are different.

After the students have decided whether to be Player 1 or Player 2 and have tossed the chips many times, they might want to revise their choices. When the class has completed the experiment and discussed the results, the theoretical probability can be analyzed by completing a tree diagram (fig. 11.1).

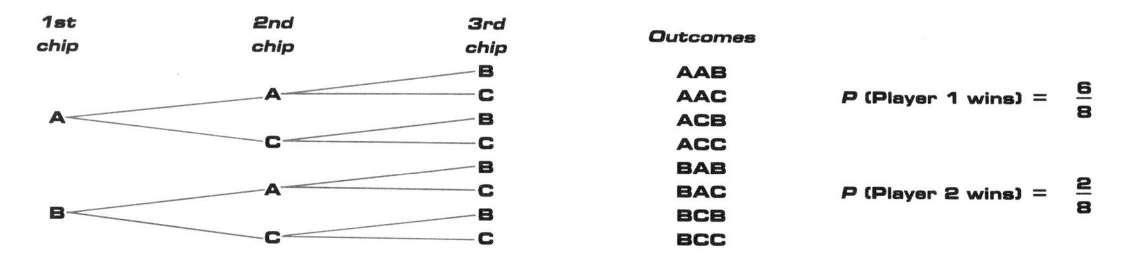

***Fig. 11.1.* Tree diagram**

Player 1 can win in six ways, but Player 2 can win in only two ways; hence, the probability of Player 1 winning is 6/8 and the game is clearly unfair.

Students should also understand that some probability problems do not have theoretical solutions. Given a set of thumbtacks, what is the probability that one thumbtack will land "point up" when tossed? Many students will guess that the answer is ½. Through experimentation, they will discover that the probability changes with the size of the head of the tack and the length of the shank.

At the K–4 level, students can flip coins, use spinners, or roll number cubes to begin their study of probability. At the middle school level, such experiments should be extended so that students can determine the probabilities inherent in more complex situations using simple methods. For example, if you are making a batch of 6 cookies from a mix into which you randomly drop 10 chocolate chips, what is the probability that you will get a cookie with at least 3 chips? Students can simulate which cookies get chips by rolling a die 10 times. Each roll of the die determines which cookie gets a chip. The same type of simulation can help students determine how many boxes of cereal they should expect to purchase to receive at least 1 each of 6 types of prizes given away randomly, 1 to a box. With a computer or set of random numbers, this problem can be extended to simulate an industrial quality-control situation or to analyze the number of defective items that might occur under certain conditions on an assembly line. All these problems, experiments, and simulations can be easily studied, performed, and analyzed by middle school students.

Once students have experimented with a problem, a computer can generate hundreds or thousands of simulated results. It is important that the computer simulation follow active student exploration. This follow-up broadens students' understanding and provides them with an opportunity to observe how a greater number of trials can refine the probability model.

The nature of probability encourages a systematic and logical approach to problem solving. Throughout their experimentation and simulation, students should be making hypotheses, testing conjectures, and refining their theories on the basis of new information. Probability also can be applied to data analysis. Students can use charts, graphs, and plots to make predictions; this activity reinforces their interpretation of the information and their derivation of other useful information.

Home runs	9
Triples	2
Doubles	16
Singles	24
Walks	11
Outs	38
Total	100

***Fig. 11.2.* Softball stats**

The table in figure 11.2 gives the record for Joan Dyer's last 100 times at bat during the softball season. She is now coming up to bat. Use the data to answer the following questions:

What is the probability that Joan will get a home run?

What is the probability that she will get a hit?

How many times can she be expected to get a walk in her next 14 times at bat?

Probability connects many areas of mathematics. For example, fraction concepts play a critical role in the study of probability. Topics such as equivalent fractions, comparison of fractions, addition and multiplication of fractions, as well as whole number operations and the relationships among fractions, decimals, and percents can be reinforced through the study of probability.

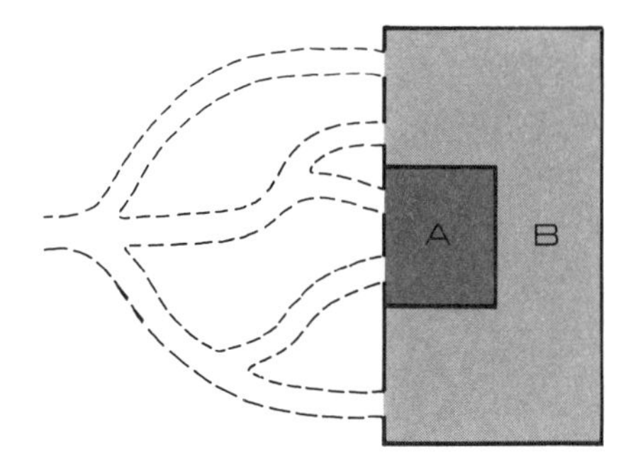

***Fig. 11.3.* Maze**

Dividing the area of a rectangle into fractional parts to model a probability problem provides an excellent opportunity for students to identify the relationship between concepts in geometry and operations with fractions.

In the maze in figure 11.3, Tom is to pick a path at random. Use the grid in figure 11.4 to determine the probability that he will enter room A or room B.

Paths	
upper	B
middle	B
	A
lower	A
	B

***Fig. 11.4.* Grid**

To calculate the probabilities directly, students would have to multiply probabilities and then add the results. However, the use of an area model makes the problem more accessible. By using fractional parts of a region, a student can represent the sequence of possible choices for paths and then add simple fractions to determine the probabilities. $P(A) = \frac{1}{6} + \frac{1}{6} = \frac{2}{6}$.

The study of probability in grades 5–8 actively engages students in exploring events and situations relevant to their daily lives.

STANDARD 12: GEOMETRY

In grades 5–8, the mathematics curriculum should include the study of the geometry of one, two, and three dimensions in a variety of situations so that students can—

- ***identify, describe, compare, and classify geometric figures;***
- ***visualize and represent geometric figures with special attention to developing spatial sense;***
- ***explore transformations of geometric figures;***
- ***represent and solve problems using geometric models;***
- ***understand and apply geometric properties and relationships;***
- ***develop an appreciation of geometry as a means of describing the physical world.***

Focus

> Geometry is grasping space . . . that space in which the child lives, breathes and moves. The space that the child must learn to know, explore, conquer, in order to live, breathe and move better in it. (Freudenthal 1973, p. 403).

The study of geometry helps students represent and make sense of the world. Geometric models provide a perspective from which students can analyze and solve problems, and geometric interpretations can help make an abstract (symbolic) representation more easily understood. Many ideas about number and measurement arise from attempts to quantify real-world objects that can be viewed geometrically. For example, the use of area models provides an interpretation for much of the arithmetic of decimals, fractions, ratios, proportions, and percents.

Students discover relationships and develop spatial sense by constructing, drawing, measuring, visualizing, comparing, transforming, and classifying geometric figures. Discussing ideas, conjecturing, and testing hypotheses precede the development of more formal summary statements. In the process, definitions become meaningful, relationships among figures are understood, and students are prepared to use these ideas to develop informal arguments. The informal exploration of geometry can be exciting and mathematically productive for middle school students. At this level, geometry should focus on investigating and using geometric ideas and relationships rather than on memorizing definitions and formulas.

The study of geometry in grades 5–8 links the informal explorations begun in grades K–4 to the more formalized processes studied in grades 9–12. The expanding logical capabilities of students in grades 5–8 allow them to draw inferences and make logical deductions from geometric problem situations. This does not imply that the study of geometry in grades 5–8 should be a formalized endeavor; rather, it should simply provide increased opportunities for students to engage in more systematic explorations.

Discussion

A teacher's questioning techniques and language in directing students' thinking are critical to the students' development of an understanding of geometric relationships. Students should be challenged to analyze their thought processes and explanations. They should be allowed sufficient time to discuss the quality of their answers and to ponder such questions as, Could it be another way? What would happen if . . .? Students should learn to use correct vocabulary, including such common terms as *and*, *or*, *all*, *some*, *always*, *never*, and *if . . . then*, to reason, as well as such words as *parallel*, *perpendicular*, and *similar* to describe. Geometry also has a vocabulary of its own, including terms like *rhombus, trapezoid*, and *dodecahedron*, and students need ample time to develop confidence in their use of this new and unique language. Definitions should evolve from experiences in constructing, visualizing, drawing, and measuring two- and three-dimensional figures, relating properties to figures, and contrasting and classifying figures according to their properties. Students who are asked to memorize a definition and a textbook example or two are unlikely to remember the term or its application.

Triangles are a subject of study in all grades, K–12. At the middle school level, most of the basic properties of triangles can be developed through investigations such as the following.

You are given a pile of toothpicks all the same size. First, take three toothpicks. Can you form a triangle using all three toothpicks placed end to end in the same plane? Can a different triangle be formed? What kinds of triangles are possible?

Now take four toothpicks and repeat the questions. Then repeat with five toothpicks, six toothpicks, and so on.

A table such as that in figure 12.1 helps students to organize their data in a systematic manner.

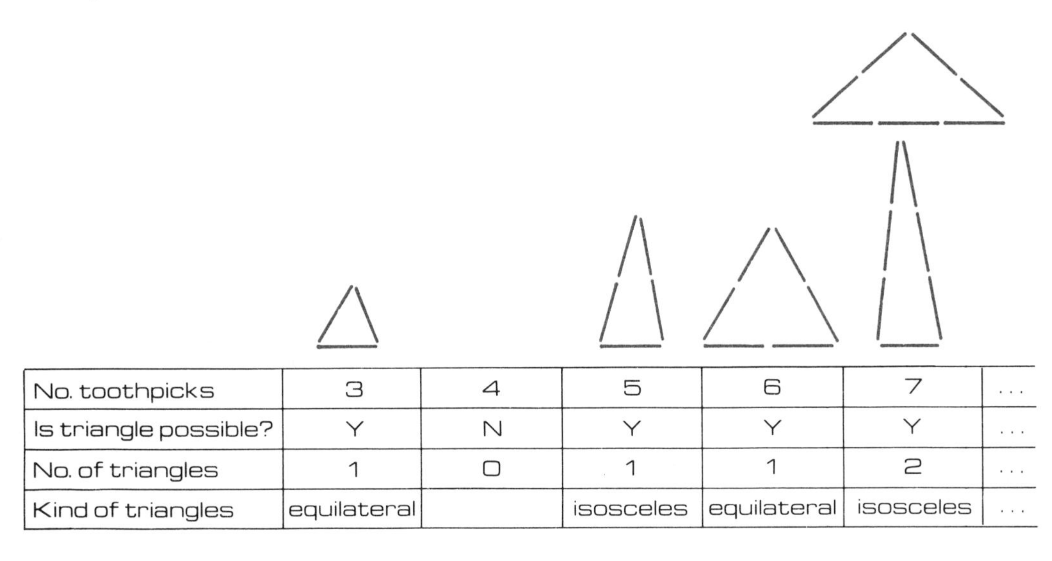

No. toothpicks	3	4	5	6	7	. . .
Is triangle possible?	Y	N	Y	Y	Y	. . .
No. of triangles	1	0	1	1	2	. . .
Kind of triangles	equilateral		isosceles	equilateral	isosceles	. . .

***Fig. 12.1.* Triangles**

In this activity, students find that the sum of the measures of two sides of a triangle must be greater than the measure of the third side. This activity also reinforces the classification of triangles by sides and angles.

One of the most important properties in geometry, the Pythagorean theorem, is introduced in the middle grades. Students can discover this relationship through explorations, such as the one suggested in figure 12.2.

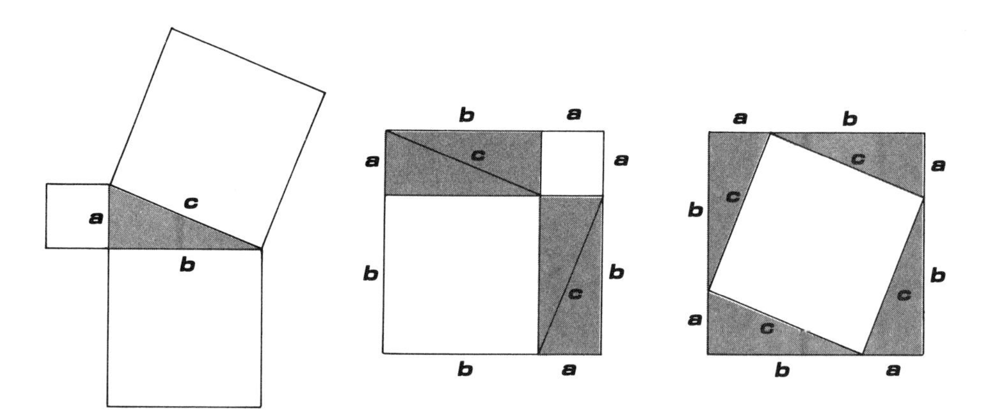

Fig. 12.2. Pythagorean theorem

Another interesting problem in which students at different levels can investigate geometric properties and relationships of quadrilaterals is shown in figure 12.3. Students can explore what happens when they connect the midpoints of the sides of several quadrilaterals. Their discovery that a parallelogram is formed can prompt them to ask such questions as, How does the area of the new figure compare to that of the quadrilateral? What quadrilateral would you start with so that the new figure is a rectangle? A square? Computer software that allows students to construct geometric figures and determine the measures of arcs, angles, and segments creates a rich environment for the investigation of geometric properties and relationships. Students can make conjectures and explore other figures to verify their reasoning.

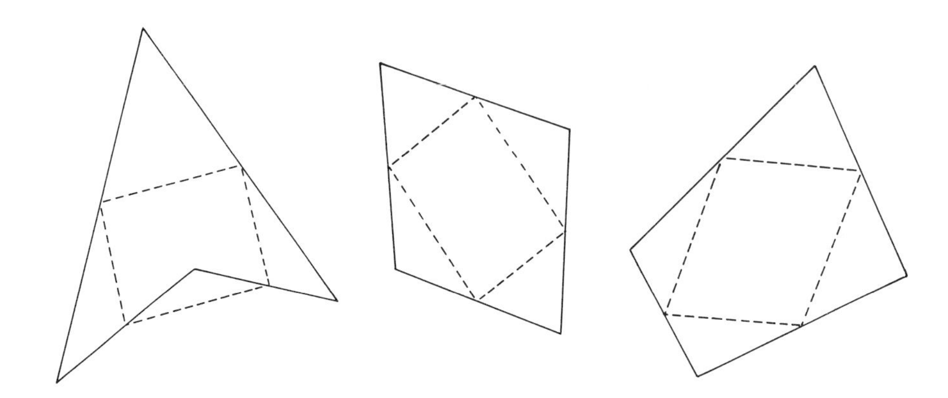

Fig. 12.3. Quadrilaterals

Computer software allows students to construct two- and three-dimensional shapes on a screen and then flip, turn, or slide them to view them from a new perspective. Explorations of flips, slides, turns, stretchers, and shrinkers will illuminate the concepts of congruence and similarity. Observing and learning to represent two- and three-dimensional figures in various positions by drawing and construction also helps students develop spatial sense.

Measuring and comparing the sides and angles of similar polygons help students develop and understand the mathematical concept of similar figures. The relationship between the angles and the sides of similar triangles is the foundation of trigonometry. Similarity also can be related to such real-world contexts as photographs, models, projections of pictures, and photocopy machines. Students should explore the relationships among the lengths, areas, and volumes of similar solids. Most students in grades 5–8 incorrectly believe that if the sides of a figure are doubled to produce a similar figure, the area and volume also will be doubled. See figure 12.4.

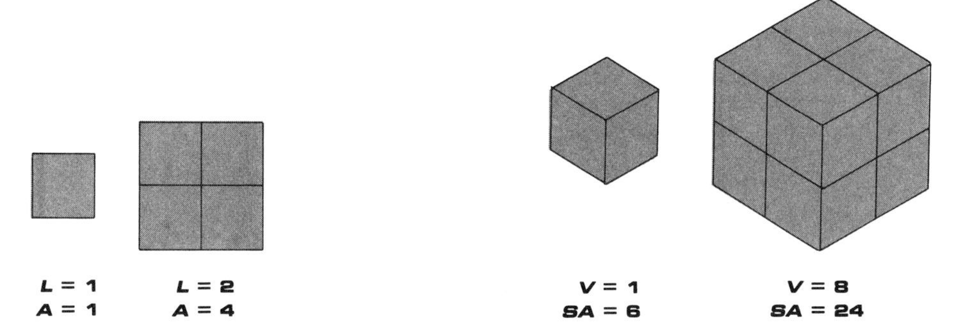

***Fig. 12.4.* Area and volume**

Investigations of two- and three-dimensional models fosters an understanding of the different growth rates for linear measures, areas, and volumes of similar figures. These ideas are fundamental to measurement and critical to scientific applications.

Students' understanding of the angle properties of polygons and the concept of area can be enhanced through explorations of tessellations with regular polygons. Which polygons will cover the plane and which ones will not? Why? This exercise can be extended to combining regular polygons and investigating solids constructed from regular polygons. In such a discussion students can also consider why the square is used as a unit of area and the cube as a unit of volume.

Symmetry in two and three dimensions provides rich opportunities for students to see geometry in the world of art, nature, construction, and so on. Butterflies, faces, flowers, arrangements of windows, reflections in water, and some pottery designs involve symmetry. Turning symmetry is illustrated by bicycle gears. Pattern symmetry can be observed in the multiplication table, in numbers arrayed in charts, and in Pascal's triangle.

Experience with geometry at the 5–8 level should sensitize students to looking at the world around them in a more meaningful way.

STANDARD 13: MEASUREMENT

In grades 5–8, the mathematics curriculum should include extensive concrete experiences using measurement so that students can—

- ***extend their understanding of the process of measurement;***
- ***estimate, make, and use measurements to describe and compare phenomena;***
- ***select appropriate units and tools to measure to the degree of accuracy required in a particular situation;***
- ***understand the structure and use of systems of measurement;***
- ***extend their understanding of the concepts of perimeter, area, volume, angle measure, capacity, and weight and mass;***
- ***develop the concepts of rates and other derived and indirect measurements;***
- ***develop formulas and procedures for determining measures to solve problems.***

Focus

Measurement activities can and should require a dynamic interaction between students and their environment. Students encounter measurement ideas both in and out of school, in such areas as architecture, art, science, commercial design, sports, cooking, shopping, and map reading. The study of measurement shows the usefulness and practical applications of mathematics, and students' need to communicate about various measurements highlights the importance of standard units and common measurement systems.

Measurement in grades 5–8 should be an active exploration of the real world. As students acquire the ability to use appropriate tools in measuring objects, they should extend these skills to new situations and new applications. The approximate nature of measure is an aspect of number that deserves repeated attention at this level. However, measurement activities in these grades should focus on using concepts and skills to solve problems and investigate other mathematical situations.

The development of the concepts of perimeter, area, volume, angle measure, capacity, and weight is initiated in grades K–4 and extended and applied in grades 5–8. At this level, students can begin to estimate the error of a measurement, adding to the K–4 notion of "about" 4 cm. From their explorations, students should develop multiplicative procedures and formulas for determining measures. The curriculum should focus on the development of understanding, not on the rote memorization of formulas. In addition, the concepts of rate as a measure and of indirect measurement are developed in grades 5–8.

Geometry and measurement are interconnected and support each other in many ways. The concept of similarity, for example, can be used in indirect measurement, and the perimeter and area of irregular figures can be determined using line segments and squares, respectively. Measurement also has strong connections to the students' expanding concept of number. Fractions, decimals, and rational numbers are used to represent measures.

Discussion

In everyday life, people need to make many kinds of measures to resolve common questions: About how long will it take? About how much do I need to buy? About how much will it hold? An estimate is often sufficient. Estimation requires a judgment about an entity's approximate relationship to a standard. Students' skills at estimating measurements will develop only through experience. One important aspect of estimating measurements is context. Students need to develop estimation strategies, and they need experience in judging what degree of accuracy is required in a given situation. If a person is buying carpet, error should be in the direction of an overestimate. However, if one is estimating how much time to sunbathe without burning, an underestimate is best. In developing estimation skills for measurement, a student learns to relate the world to familiar personal experiences. The ability to hold one's hands about a meter apart, to know the length of a foot or stride, to know the width of a fingernail—all these are useful estimating tools.

During students' early experiences with counting and operations using whole numbers, they work with precise situations that yield exact counts. Measuring the length of an object is quite different, and it is essential that students understand this difference. The approximate nature of measuring is a concept that takes time and many experiences for students to develop and understand. The following classroom activity helps students with this concept and relates to the standard on statistics.

Have each student use a meter tape to measure the length of the room to the nearest centimeter. Record each student's measure and analyze the results (see fig. 13.1).

95|4

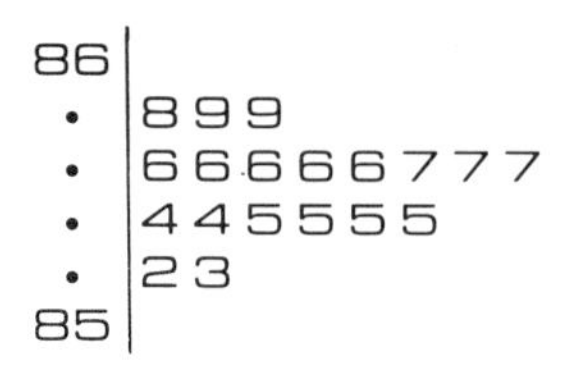

85|3 means 853 centimeters

Stem and Leaf

852 853 854 855 856 857 858 859 860 954

Box and Whiskers

***Fig. 13.1.* Data plots**

Linked to the development of measuring concepts are experiences with standard measuring tools: rulers, balances, protractors, clocks, wheels, speedometers, and so on. In a given situation, a student must select both an appropriate unit and a tool to find a measurement; this selection depends on the degree of accuracy required in a particular situation. It would be inappropriate to select a 10-cm ruler to measure the length of a soccer field, even when a fairly accurate measure is needed. However, the square corner of a sheet of paper can be used to "measure" an angle if one only needs to know whether it is larger or smaller than a right angle.

As students progress through grades 5–8, they should develop more efficient procedures and, ultimately, formulas for finding measures. Length, area, and volume of one-, two-, and three-dimensional figures are especially important over these grade levels. For example, once students have discovered that it is possible to find the area of a rectangle by covering a figure with squares and then counting, they are ready to explore the relationship between areas of rectangles and areas of other geometric figures. This exploration gives students an opportunity to reason deductively and see how mathematical ideas relate to one another. The following sketches suggest some possibilities.

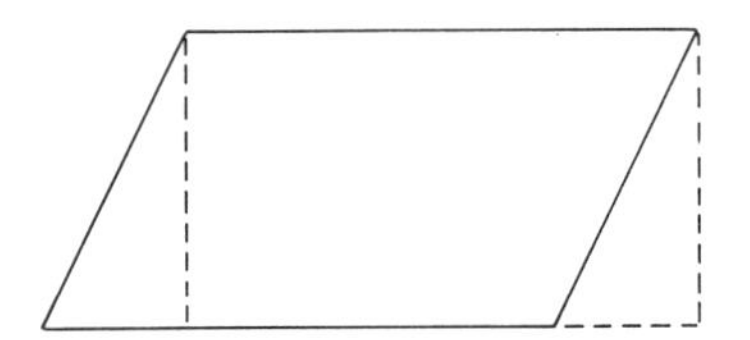

***Fig. 13.2.* Parallelogram to rectangle**

The area of a parallelogram can be rearranged into a rectangle (fig. 13.2).

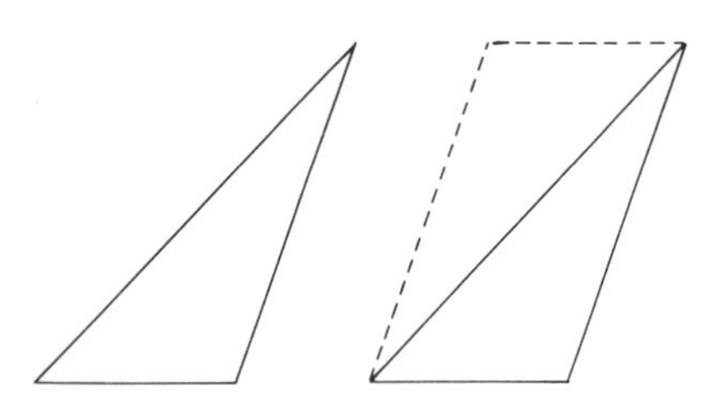

***Fig. 13.3.* Triangle to parallelogram**

The area of a triangle is one-half the area of a parallelogram (fig. 13.3).

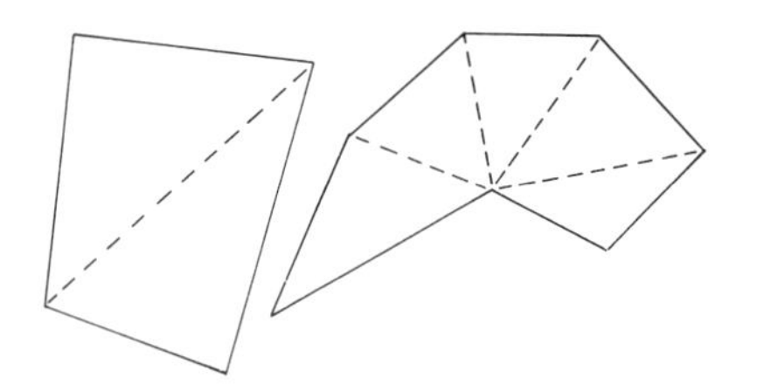

***Fig. 13.4.* Partitioning polygons into triangles**

All polygons can be partitioned into triangles (fig. 13.4).

All these connections require students to understand that the area of a figure does not change if it is partitioned and rearranged. It is also important that students understand the association between multiplication and determining the area of a rectangle. The formula is not a "magic box." It is a summary of a process that tells how many units it takes to cover the rectangle. It is also a summary of the relationship among area, height, and length. Any two of these determine the other: $A = LH$; $L = A/H$; $H = A/L$.

An example of a practical problem that involves measurement, similarity in scale drawing, and creativity is the following (Wirszup and Streit 1987).

Given a piece of plywood 150 cm x 300 cm, design a dog kennel that can be made from the piece. Try to make your kennel as large as you can. Make a scale drawing to show how the parts of the kennel have to be cut from the plywood. Give the measurements. Draw a sketch or sketches to show what the finished kennel will look like. Write the measurements on the sketches.

Students need many experiences with the concepts of rate in measurement settings. Here is an example of a problem that uses rates as measures (Meyer and Sallee 1983):

It is the seventh annual cross-country motorcycle race across the Nevada desert, 70 miles and back. Orite, on her new Harley-Davidson, averages 80 miles an hour going out but has clutch trouble and can manage only 60 miles an hour coming back. Eric, on a Honda, can go only 70 miles an hour, but he keeps it up for the entire race. Who wins the race?

Planets	Distance from sun (in millions of miles)
Mercury	36
Venus	67
Earth	93
Mars	142
Jupiter	484
Saturn	884
Uranus	1789
Neptune	2809
Pluto	3685

***Fig. 13.5.* Solar system**

Constructing a scale model of the solar system is another problem that involves proportional reasoning and connects mathematics to another discipline. The gymnasium or the hall of the school can be used. Students have to decide what will represent the orbit of Pluto and then figure out what the radius of the other orbits will be. See figure 13.5.

Areas of irregular figures can be approximated by covering the figure with a square grid and counting the whole squares within the figure as an inner measure and all squares that touch the figure anywhere as an outer measure. The actual measure is between these two, so the mean of the measures gives an estimate of the area and half the difference between the measures gives the greatest possible error. If the possible error is too big, the process can be repeated with a smaller grid. See figure 13.6.

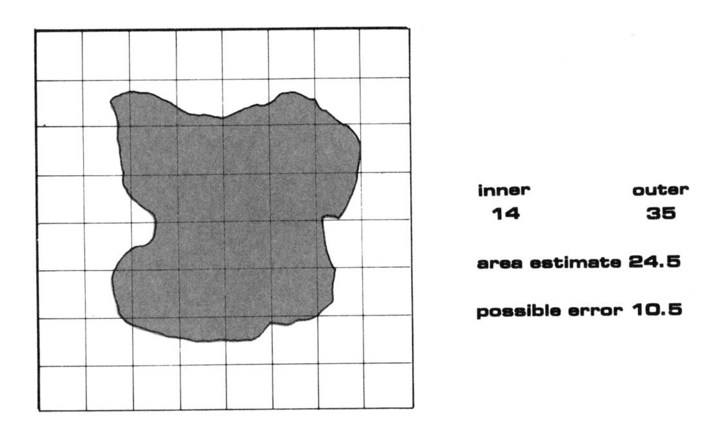

Fig. 13.6. **Estimating area**

Students can use their knowledge of similar triangles to measure heights of inaccessible objects. Two possibilities are illustrated in figure 13.7, one using shadows and the other using reflections in a mirror.

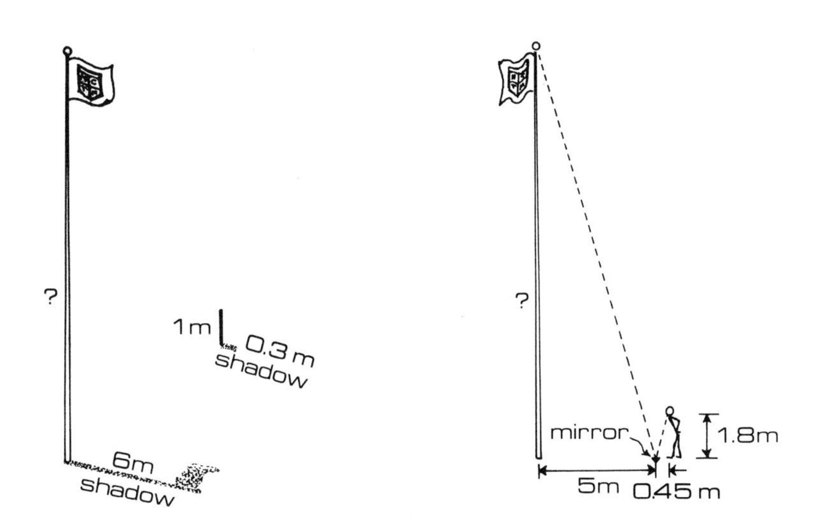

Fig. 13.7. **Indirect measurement**

Measurement experiences are a powerful mathematical connection among topics in the middle school curriculum and in other disciplines. Measurement clearly shows the usefulness of mathematics in everyday life.

GRADES 9—12

♦ ♦ ♦ ♦ ♦ ♦ ♦ ♦

CURRICULUM STANDARDS FOR GRADES 9–12

OVERVIEW

This section presents fourteen curriculum standards for grades 9–12:

1. ***Mathematics as Problem Solving***
2. ***Mathematics as Communication***
3. ***Mathematics as Reasoning***
4. ***Mathematical Connections***
5. ***Algebra***
6. ***Functions***
7. ***Geometry from a Synthetic Perspective***
8. ***Geometry from an Algebraic Perspective***
9. ***Trigonometry***
10. ***Statistics***
11. ***Probability***
12. ***Discrete Mathematics***
13. ***Conceptual Underpinnings of Calculus***
14. ***Mathematical Structure***

Background

Historically, the purposes of secondary school mathematics have been to provide students with opportunities to acquire the mathematical knowledge, skills, and modes of thought needed for daily life and effective citizenship, to prepare students for occupations that do not require formal study after graduation, and to prepare students for postsecondary education, particularly college. The *Standards'* Introduction describes a vision of school mathematics in which these purposes are embedded in a context that is both broader and more consistent with accelerating changes in today's society. High school graduates during the remainder of this century can expect to have four or more career changes. To develop the requisite adaptability, high school mathematics instruction must adopt broader goals for *all* students. It must provide experiences that encourage and enable students to value mathematics, gain confidence in their own mathematical ability, become mathematical problem solvers, communicate mathematically, and reason mathematically. The fourteen standards for grades 9–12 establish a framework for a core curriculum that reflects the needs of all students, explicitly recognizing that they will spend their adult lives in a society increasingly dominated by technology and quantitative methods.

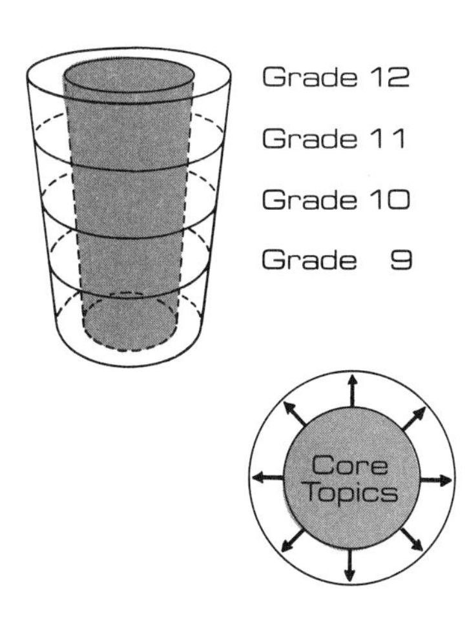

Fig. 1. A differentiated core curriculum

In view of existing disparities in educational opportunity in mathematics and the increasing necessity that all individuals have options for further education and alternative careers, each standard identifies the mathematical content or processes and the associated student activities that should be included in the curriculum for *all* students. As suggested by figure 1, the core curriculum is intended to provide a common body of mathematical ideas accessible to all students. We recognize that students entering high school differ in many ways, including mathematical achievement, but we believe these differences are best addressed by enrichment and extensions of the proposed content rather than by deletions. The mathematics curriculum must set high, but reasonable, expectations for *all* students.

The core curriculum can be extended in a variety of ways to meet the needs, interests, and performance levels of individual students or groups of students. To illustrate, many of the standards also specify topics that should be studied by college-intending students. We use the term *college-intending* not in an exclusionary sense, but only as a means by which to identify the additional mathematical topics that should be studied by students who plan to attend college. In fact, we believe that these additional curricular topics should be studied by all students who have demonstrated interest and achievement in mathematics.

A school curriculum in line with these standards should be organized so as to permit all students to progress as far into the mathematics proposed here as their achievement with the topic allows. In particular, students with exceptional mathematical talent who advance through the material more quickly than others may continue to college-level work in the mathematical sciences. However, we strongly recommend against acceleration that either omits content identified in these standards or advances students through it superficially.

Figure 1 also is intended to portray an expectation that mathematical ideas will grow and deepen as students progress through the curriculum and that the consolidation of learning is essential for all students during the senior year. Such a synthesis of mathematical knowledge will enhance students' prospects for securing employment and for both entering and successfully completing collegiate programs. It is, therefore, an underpinning of the proposed curriculum.

Underlying Assumptions

The standards for grades 9–12 are based on the following assumptions:

- Students entering grade 9 will have experienced mathematics in the context of the broad, rich curriculum outlined in the K–8 standards.
- The level of computational proficiency suggested in the K–8 standards will be expected of all students; however, no student will be denied access to the study of mathematics in grades 9–12 because of a lack of computational facility.
- Although arithmetic computation will not be a direct object of study in grades 9–12, number and operation sense, estimation skills, and the ability to judge the reasonableness of results will be strengthened in the context of applications and problem solving, including those situations dealing with issues of scientific computation.
- Scientific calculators with graphing capabilities will be available to all students at all times.
- A computer will be available at all times in every classroom for demonstration purposes, and all students will have access to computers for individual and group work.
- At least three years of mathematical study will be required of all secondary school students.

- These three years of mathematical study will revolve around a core curriculum differentiated by the depth and breadth of the treatment of topics and by the nature of applications.

- Four years of mathematical study will be required of all college-intending students.

- These four years of mathematical study will revolve around a broadened curriculum that includes extensions of the core topics and for which calculus is no longer viewed as *the* capstone experience.

- All students will study appropriate mathematics during their senior year.

Features of the Mathematics Content

Initially, it may appear that an excessive amount of curriculum content is described in the 9–12 standards. When this content is evaluated, however, it should be remembered that the proposed 5–8 curriculum will enable students to enter high school with substantial gains in their conceptual and procedural understandings of algebra, in their knowledge of geometric concepts and relationships, and in their familiarity with informal, but conceptually based, methods for dealing with data and situations involving uncertainty. Moreover, additional instructional time can be gained by organizing the curriculum so that student learning is systematically maintained and review is embedded in the context of new topics or problem situations. With these conditions satisfied, it is our belief that it will be possible to address the recommended content within a three- or four-year sequence with the expectation of a reasonable level of student proficiency.

Traditional topics of algebra, geometry, trigonometry, and functions remain important components of the secondary school mathematics curriculum. However, the 9–12 standards call for a shift in emphasis from a curriculum dominated by memorization of isolated facts and procedures and by proficiency with paper-and-pencil skills to one that emphasizes conceptual understandings, multiple representations and connections, mathematical modeling, and mathematical problem solving. The integration of ideas from algebra and geometry is particularly strong, with graphical representation playing an important connecting role. Thus, frequent reference to *graphing utilities* will be found throughout these standards; by this we mean a computer with appropriate graphing software or a graphing calculator. In addition, topics from statistics, probability, and discrete mathematics are elevated to a more central position in the curriculum for all students. Specific topics that should be given either increased or reduced emphasis are summarized in the chart on pages 126–127.

Patterns of Instruction

The broadened view of mathematics described in the Introduction to this document under the rubric *mathematical power*, together with the capabilities of available and emerging technology, suggests a need for changes in instructional patterns and in the roles of both teachers and students.

A variety of instructional methods should be used in classrooms in order to cultivate students' abilities to investigate, to make sense of, and to construct meanings from new situations; to make and provide arguments for conjectures; and to use a flexible set of strategies to solve problems

♦ ♦ ♦ ♦ ♦ ♦ ♦ ♦

SUMMARY OF CHANGES IN CONTENT

TOPICS TO RECEIVE INCREASED ATTENTION

ALGEBRA
- The use of real-world problems to motivate and apply theory
- The use of computer utilities to develop conceptual understanding
- Computer-based methods such as successive approximations and graphing utilities for solving equations and inequalities
- The structure of number systems
- Matrices and their applications

GEOMETRY
- Integration across topics at all grade levels
- Coordinate and transformation approaches
- The development of short sequences of theorems
- Deductive arguments expressed orally and in sentence or paragraph form
- Computer-based explorations of 2-D and 3-D figures
- Three-dimensional geometry
- Real-world applications and modeling

TRIGONOMETRY
- The use of appropriate scientific calculators
- Realistic applications and modeling
- Connections among the right triangle ratios, trigonometric functions, and circular functions
- The use of graphing utilities for solving equations and inequalities

FUNCTIONS
- Integration across topics at all grade levels
- The connections among a problem situation, its model as a function in symbolic form, and the graph of that function
- Function equations expressed in standardized form as checks on the reasonableness of graphs produced by graphing utilities
- Functions that are constructed as models of real-world problems

STATISTICS

PROBABILITY

DISCRETE MATHEMATICS

♦ ♦ ♦ ♦ ♦ ♦ ♦ ♦

AND EMPHASES IN 9–12 MATHEMATICS

TOPICS TO RECEIVE DECREASED ATTENTION

ALGEBRA
- Word problems by type, such as coin, digit, and work
- The simplification of radical expressions
- The use of factoring to solve equations and to simplify rational expressions
- Operations with rational expressions
- Paper-and-pencil graphing of equations by point plotting
- Logarithm calculations using tables and interpolation
- The solution of systems of equations using determinants
- Conic sections

GEOMETRY
- Euclidean geometry as a complete axiomatic system
- Proofs of incidence and betweenness theorems
- Geometry from a synthetic viewpoint
- Two-column proofs
- Inscribed and circumscribed polygons
- Theorems for circles involving segment ratios
- Analytic geometry as a separate course

TRIGONOMETRY
- The verification of complex identities
- Numerical applications of sum, difference, double-angle, and half-angle identities
- Calculations using tables and interpolation
- Paper-and-pencil solutions of trigonometric equations

FUNCTIONS
- Paper-and-pencil evaluation
- The graphing of functions by hand using tables of values
- Formulas given as models of real-world problems
- The expression of function equations in standardized form in order to graph them
- Treatment as a separate course

from both within and outside mathematics. In addition to traditional teacher demonstrations and teacher-led discussions, greater opportunities should be provided for small-group work, individual explorations, peer instruction, and whole-class discussions in which the teacher serves as a moderator.

These alternative methods of instruction will require the teacher's role to shift from dispensing information to facilitating learning, from that of director to that of catalyst and coach. The introduction of new topics and most subsumed objectives should, whenever possible, be embedded in problem situations posed in an environment that encourages students to explore, formulate and test conjectures, prove generalizations, and discuss and apply the results of their investigations. Such an instructional setting enables students to approach the learning of mathematics both creatively and independently and thereby strengthen their confidence and skill in doing mathematics.

The role of students in the learning process in grades 9–12 should shift in preparation for their entrance into the work force or higher education. Experiences designed to foster continued intellectual curiosity and increasing independence should encourage students to become self-directed learners who routinely engage in constructing, symbolizing, applying, and generalizing mathematical ideas. Such experiences are essential in order for students to develop the capability for their own lifelong learning and to internalize the view that mathematics is a process, a body of knowledge, and a human creation.

The use of technology in instruction should further alter both the teaching and the learning of mathematics. Computer software can be used effectively for class demonstrations and independently by students to explore additional examples, perform independent investigations, generate and summarize data as part of a project, or complete assignments. Calculators and computers with appropriate software transform the mathematics classroom into a laboratory much like the environment in many science classes, where students use technology to investigate, conjecture, and verify their findings. In this setting, the teacher encourages experimentation and provides opportunities for students to summarize ideas and establish connections with previously studied topics.

The most fundamental consequence of changes in patterns of instruction in response to technology-rich classroom environments is the emergence of a new classroom dynamic in which teachers and students become natural partners in developing mathematical ideas and solving mathematical problems.

Assessment of student learning should be viewed as an integral part of instruction and should be aligned with key aspects of instruction, such as the use of technology. The reader is encouraged to examine the Evaluation standards for more detail on student assessment.

The following chart summarizes the major changes in patterns of instruction proposed for grades 9–12.

♦ ♦ ♦ ♦ ♦ ♦ ♦ ♦

SUMMARY OF CHANGES IN INSTRUCTIONAL PRACTICES IN 9–12 MATHEMATICS

INCREASED ATTENTION to—

- The active involvement of students in constructing and applying mathematical ideas
- Problem solving as a means as well as a goal of instruction
- Effective questioning techniques that promote student interaction
- The use of a variety of instructional formats (small groups, individual explorations, peer instruction, whole-class discussions, project work)
- The use of calculators and computers as tools for learning and doing mathematics
- Student communication of mathematical ideas orally and in writing
- The establishment and application of the interrelatedness of mathematical topics
- The systematic maintenance of student learnings and embedding review in the context of new topics and problem situations
- The assessment of learning as an integral part of instruction

DECREASED ATTENTION to—

- Teacher and text as exclusive sources of knowledge
- Rote memorization of facts and procedures
- Extended periods of individual seatwork practicing routine tasks
- Instruction by teacher exposition
- Paper-and-pencil manipulative skill work
- The relegation of testing to an adjunct role with the sole purpose of assigning grades

The Core Curriculum

The core curriculum for all students is the most fundamental change proposed for grades 9–12. It is very important for the reader of these standards to understand (1) exactly what is and is not being proposed, (2) the advantages of the core curriculum over current practice, and (3) the implications for teaching as decision making.

What is and what is not being proposed. For emphasis, we repeat the core-curriculum assumption here (slightly modified to highlight its significant characteristics):

> [These] three years of [required] mathematical study will revolve around a core curriculum differentiated by the depth and breadth of the treatment of topics and by the nature of applications.

This assumption proposes that the *curriculum topics* described in this document apply to *all* students—except where the topics are specifically differentiated for those who are college intending. This means that the longstanding practice of requiring lower-achieving high school students to repeat sixth-grade mathematics content over and over will be replaced by a study of content that we believe provides these students, as well as their classmates, with a central core of mathematical representation, mathematical processing, mathematical problem solving, and mathematical thinking. In particular, this statement asserts that if the sequence of courses often designated as "general mathematics" does not address the content and associated goals of the core curriculum, it is no longer acceptable.

It is important to understand that this statement does *not* imply that students of all performance levels must be taught in the same classroom, and it does *not* imply that the content presentation for all students must be the same. However, no matter how individual school districts or schools choose to deal organizationally with students who exhibit different abilities, achievement levels, and interests, it is crucial that all students experience the full range of topics proposed for the core curriculum. It is equally important to ensure that the instructional practices and resources described in the previous section are integral to the mathematical experiences of all students.

Advantages this proposal offers to both students and teachers. The core curriculum proposed here offers several advantages over current practice, which dichotomizes secondary school mathematics into programs for college-bound and non-college-bound students.

Advantage 1: The core curriculum provides equal access and opportunity to all students. Mathematical literacy is vital to every individual's meaningful and productive life. The mathematical abilities needed for everyday life and for effective citizenship have changed dramatically over the last decade and are no longer provided by a computation-based general mathematics program. By removing the "computational gate" to the study of high school mathematics and recognizing that there frequently is not a strict hierarchy among the proposed mathematics topics at this level, we are able to afford all students more opportunities to fulfill their mathematical potential and participate throughout their lives as productive members of our society.

Advantage 2: The core curriculum provides greater flexibility for individual students, thus allowing them to keep their options open. Students' interests, goals, and achievements are not static but change as they mature and advance through high school. In choosing not to trap students in one of the two conventional linear patterns, we ensure that doors to college programs and vocational training are kept open for all students.

Advantage 3: The core curriculum better prepares non-college-intending students for the world of today and tomorrow. Henry Pollak's summary of the types of expectations stated by today's employers (reported on page 4 of this document), coupled with the report *Workforce 2000: Work and Workers for the Twenty-first Century* (Johnston and Packer 1987), makes it apparent that much higher mathematics, language, and reasoning capabilities will be required of employees. The ever-increasing role of technology in our society further argues for a curriculum that moves all students beyond computation.

Advantage 4: The core curriculum provides opportunities for all students to confront more interesting and important mathematics. By assigning computational algorithms to calculator or computer processing, this curriculum seeks not only to move students forward but to capture their interest. As a result, students no longer will be confronted with the demeaning prospect of studying for the third, fourth, or fifth time the same content topics as their twelve-year-old siblings. We believe the opportunity to study mathematics that is more interesting and useful and not characterized as remedial will enhance students' self-concepts as well as their attitudes toward, and interest in, mathematics. These attitudinal shifts, coupled with the changes in mathematical content, should in turn provide a more interesting and stimulating environment for teaching.

In summary, the core curriculum seeks to provide a fresh approach to mathematics for all students—one that builds on what students can do rather than on what they cannot do.

Implications for teaching as decision making. In most schools, complete implementation of the 9–12 curriculum standards will require a transition period. Initially, it is likely that few students will have had the kind of mathematics experiences outlined in the K–8 standards, and teachers may need to provide some of the learning experiences described in the 5–8 standards as prerequisites to the proposed 9–12 curriculum. The amount of instructional time spent on informal activities will vary and will depend on the maturity of the students, their prior experience with a topic, and the complexity of the topic itself. In general, the relative maturity of high school students will enable them to progress more rapidly through certain topics than middle grade students could. Nevertheless, the importance of informal experiences for developing primitive conceptual understanding, a prerequisite to students' formal study and abstraction of mathematical ideas, should be recognized. In most situations, new mathematical ideas should continue to be introduced at the concrete level.

Once the prerequisite conditions have been met, we believe that a substantial amount of content in each topic area will be accessible to all students. This is not to imply that all students will experience the same coverage of each topic, but rather that the range of *content topics* is open to all. Often, but not always, the depth to which a topic is explored will relate to the level of abstraction at which students are capable of operating. Concrete examples and applications of a topic should be open to all; higher levels of abstraction and generalization should be available to, but not required of, all. Decisions about appropriate depth of treatment for a particular topic are a matter of teacher judgment, but we urge teachers to challenge students as much as possible. Preliminary data from other countries (Schoen 1988) developing a common curriculum for the majority of high school–aged students suggest that as teachers gain experience with these topics and the concept of content differentation, they develop effective techniques for advancing students further into the topics.

As they organize instruction for the core curriculum, it is crucial that educators differentiate between *content topic* and *content*. The *Standards* proposes that *all* students be guaranteed equal access to the same curricular *topics*; it does not suggest that all students should explore the content to the same depth or at the same level of formalism. The curricular topics we propose may each be further and quite naturally subdivided and their associated content developed at several levels, consistent with students' ability to abstract. It is at this level that differentiation takes place.

Examples of Content Differentiation

The following two examples suggest the kind of content differentiation on which the concept of the core curriculum is based. The levels indicated in these and other examples in the 9–12 standards should not be consid-

◆ ◆ ◆ ◆ ◆ ◆ ◆ ◆

ered ceilings; students of varying abilities and differing needs should be encouraged and helped to progress to as high a level as they demonstrate the interest and capacity to understand. The differentiation of content, if well planned, will facilitate growth in mathematical understanding for all students.

The first example focuses on a consumer application of mathematics.

Carlos deposits $100 in a savings account earning 6% interest compounded annually. Assuming a fixed interest rate, how much money will be in the account at the end of 10 years?

Level 1: With a calculator, all students should be able to determine the amount of money in the account each year by successively applying the following relationship:

Amount at end of year = Amount at beginning of year + .06 (Amount at beginning of year).

Applying this relationship will give the following:

Amount (in dollars) at beginning of year 1 (initial deposit) = 100
Year 1: Amount at end of year = 100 + .06(100) = 106
Year 2: Amount at end of year = 106 + .06(106) = 112.36
Year 3: Amount at end of year = 112.36 + .06(112.36) = 119.10
.
.
.
Year 10: Amount at end of year = 168.95 + .06(168.95) = 179.08

The use of a computer spreadsheet would be a natural extension of this activity and would give these students a powerful tool for further processing.

Students at this level also could use a pattern or template for processing:

Year 1: Amount at end of year: $100(1.06)^1$
Year 2: Amount at end of year: $100(1.06)^2$
.
.
.
Year 10: Amount at end of year: $100(1.06)^{10}$

They could verify this template by checking its results against corresponding values obtained by direct calculation. A discussion of how this template could be modified under different initial conditions would instill in the students further mathematical power. For example:

If the time is changed to 20 years, the final amount = $100(1.06)^{20}$.
If the starting amount is $200, the final amount = $200(1.06)^{10}$.
If the annual interest rate is changed to 12%, the final amount = $100(1.12)^{10}$.

Level 2: After completing the level 1 activities, students could generalize the results of their initial year-by-year calculations and use appropriate notation to assist in this process. Except for the use of technology to aid in processing, their approach as illustrated below would be traditional.

♦ ♦ ♦ ♦ ♦ ♦ ♦ ♦

They could modify the original year-by-year table to be

$$A_0 = 100$$
$$A_1 = 100 + 100(.06) = 100(1.06)$$

or

$$A_1 = A_0 + A_0(.06) = A_0(1.06).$$

Similarly,

$$A_2 = A_1 + A_1(.06) = A_1(1.06)$$
$$A_3 = A_2 + A_2(.06) = A_2(1.06)$$
$$\vdots$$
$$A_{10} = A_9(1.06). \qquad \text{Equation(1)}$$

Combining

$$A_1 = A_0(1.06)$$
$$A_2 = A_1(1.06) = A_0(1.06)(1.06) = A_0(1.06)^2$$
$$A_3 = A_2(1.06) = A_0(1.06)^2(1.06) = A_0(1.06)^3$$
$$\vdots$$
$$A_{10} = A_0(1.06)^{10}$$

and in general, $A_n = A_0(1.06)^n$, where n is the number of years.

Further generalizing would give

$$A_n = A_0(1 + r)^n, \qquad \text{Equation (2)}$$

where A_0 is the initial deposit, r is the annual interest rate, n is the number of years, and A_n is the accumulated amount in the account after n years.

Level 3: After completing level 2, these students would further generalize equation (2) where r becomes the interest rate per interest period and n is the number of interest periods. This extension would allow students to explore problems in which the annual interest rate is compounded semiannually, quarterly, monthly, daily, and so forth.

It is important to note that all students at levels 1, 2, and 3 would use calculators or computers to address related problems such as the following and discuss their results. (1) Does the amount double if (*a*) the interest rate is doubled or (*b*) the time period is doubled? (2) What are the doubling periods for amounts invested at 7%, 10%, and 20%? (3) Some people use the rule of 72, $d = 72/100i$, to approximate the doubling period, d, for interest rate, i. Test the formula for different interest rates and report on its accuracy. (4) How much money would you need to invest today to have $10 000 in 20 years? The solution methods for these problems could involve guessing and checking, generating a table by calculator or spreadsheet, graphing, and applying the compound interest formula. In each situation, lower-achieving students would reach their conclusions through additional numerical exercises.

Level 4: After completing level 3, students would, given three of the four variables in equation (2), solve for the fourth (e.g., $r = \sqrt[n]{A_n/A_0} - 1$).

Students at this level might also explore the derivation of the rule of 72 by solving equation (2) using natural logarithms.

A variety of further extensions are possible. Equation (2) could be generalized to problems in which compound growth provides an appropriate representation (e.g., in biology). Earlier results could be proved by mathematical induction. Instantaneous compounding could be investigated, which would in turn lead to the development and use of the constant e.

As another example of how the content identified for the core curriculum can be differentiated in both the depth and formalism of treatment, consider the topic of mathematical modeling called for in the standard on problem solving. The reader should note that this particular activity would occur at a point in the curriculum somewhat later than the preceding one. Thus, a particular level in this example does not necessarily presume the same prerequisite understandings and degree of mathematical maturity as the corresponding level in the previous example.

A container manufacturing company has been contracted to design and manufacture cylindrical cans for fruit juice. The volume of each can is to be 0.946 liters. In order to minimize production costs, the company wishes to design a can that requires the smallest amount of material possible. What should the dimensions of the can be?

At each level the development of the mathematical model should include opportunities for class discussion regarding the simplification of the problem in terms of the thickness of the material and wasted material when components are cut. Next, all students would derive (as necessary) formulas for the volume ($V = \pi r^2 h$) and surface area [$S = 2\pi r(h + r)$] of a cylinder and then form cylinders of various dimensions (fig. 2), comparing their volumes and surface areas.

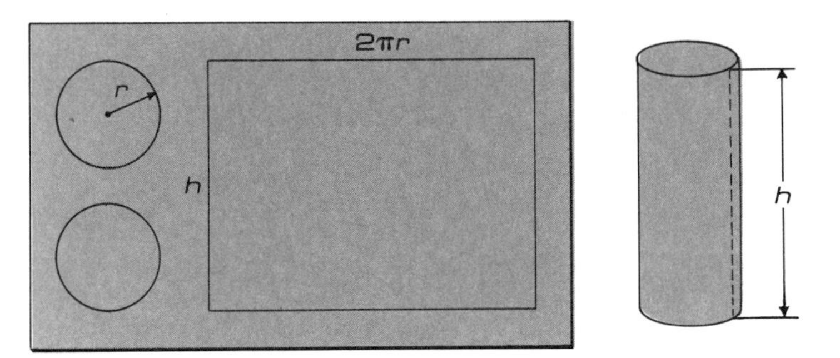

***Fig. 2.* Cylinder components**

Level 1: Students should first note that since the volume is to be 0.946 liters, or 946 ml, and since 1 ml = 1 cm^3, the dimensions of the can can be most easily expressed in centimeters. Rewriting the formula $V = 946 = \pi r^2 h$ as $h = 946/(\pi r^2)$ permits the height to be found once a value for r is specified. Now students can use a calculator to build a table of values (table 1) consisting of heights and areas corresponding to chosen radii. The table would be analyzed for an approximate solution (radius between 5 cm and 6 cm with corresponding heights between 12.0 cm and 8.4 cm).

TABLE 1
Cylinder Data

Radius (r) cm	Height (h) cm	Surface Area (S) cm^2
1	301.1	1898.3
2	75.3	971.1
3	33.3	687.2
4	18.8	573.5
5	12.0	535.5
6	8.4	541.5
7	6.1	578.2
8	4.7	638.6
9	3.7	719.2
10	3.0	817.5

♦ ♦ ♦ ♦ ♦ ♦ ♦ ♦

Level 2: Students would develop the algebraic expressions as in level 1 and design an algorithm to produce appropriate values. One possible algorithm is shown below.

For $r = 0.5, 1, 1.5, \ldots, 10$
    $h \leftarrow 946/(3.14159 \cdot r^2)$
    $S \leftarrow 2(3.14159) \cdot r(h + r)$
Output, r, h, S

Following a computer implementation of the algorithm, students would analyze the given output (table 2). This would suggest that the optimal length for the radius is again between 5 cm and 6 cm. However, the advantage of using computer methods is that students can easily approximate this length to the nearest 0.1 cm by simply modifying the program so the loop runs between 5 and 6 in increments of 0.1. A run of the modified program yields $r \doteq 5.3$ cm and $h \doteq 10.7$ cm.

TABLE 2
Refined Cylinder Data

Radius (r) cm	Height (h) cm	Surface Area (S) cm^2
0.5	1204.485 63	3785.570 8
1	301.121 407	1898.283 18
1.5	133.831 736	1275.470 49
2	75.280 351 7	971.132 72
2.5	48.179 425 1	796.069 875
3	33.457 934 1	687.215 287
3.5	24.581 339 3	617.540 384
4	18.820 087 9	573.530 88
4.5	14.870 192 9	547.678 84
5	12.044 856 3	535.479 5
5.5	9.954 426 67	534.066 195
6	8.364 483 52	541.527 813
6.5	7.127 133 88	556.541 279
7	6.145 334 83	578.161 534
7.5	5.353 269 45	605.695 542
8	4.705 021 98	638.623 52
8.5	4.167 770 33	676.547 991
9	3.717 548 23	719.159 803
9.5	3.336 525 28	766.214 89
10	3.011 214 07	817.518 001

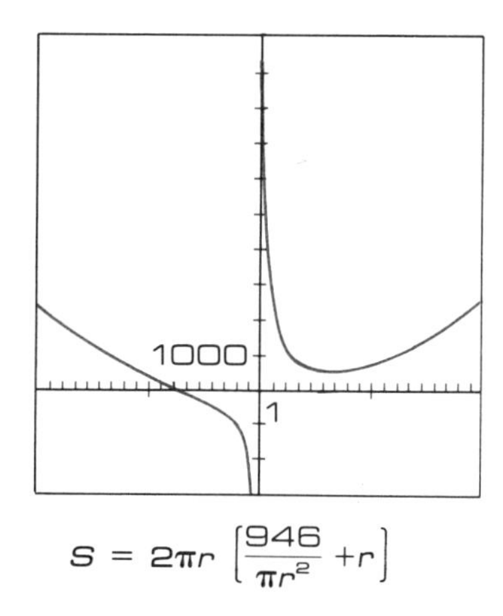

$$S = 2\pi r\left[\frac{946}{\pi r^2} + r\right]$$

Fig. 3. Graph of algebraic representation

Level 3: At this level, students would use the specified volume and the formulas for volume and surface area to express the surface area as a function of the radius. A graphing utility would be used to plot the function, and the graph (figs. 3 and 4) would be examined to determine the minimum surface area, S, and corresponding radius, r, from which the corresponding height, h, can be determined.

Level 4: After completing either the level 2 or level 3 activity, students would use different volume values in an attempt to discover a general relationship between radius and height that would minimize the surface area for a given volume. The analysis of accumulated data from several trials will suggest that the surface area appears to be minimal when h is twice r.

Level 5: Students at this level would prove that the surface area is minimal when $h = 2r$. This task should be viewed as an open-ended project, since it would not only require the generalization of the more familiar arithmetic-geometric mean inequality for two numbers to that for three numbers $[(a + b + c)/3 \geqslant \sqrt[3]{abc}$ and the equality holds if and only if $a = b = c]$ but would also entail ingenuity and creative mathematical thought.

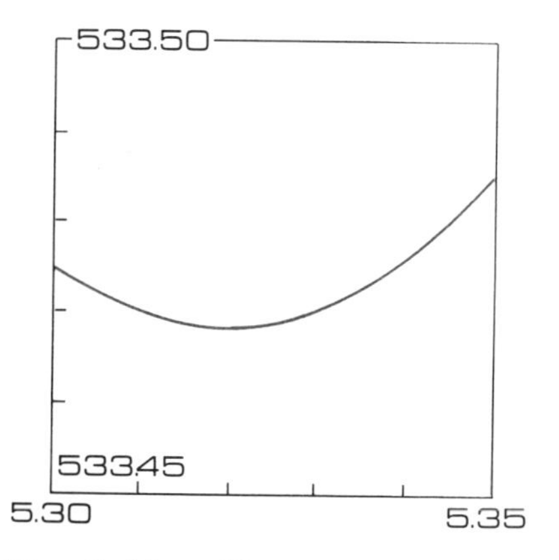

Fig. 4. Magnification showing that the minimum value of S is about 533.47 when r is about 5.32 (in this instance, $h \doteq$ 10.64)

One approach is to observe that since

$$S = 2\pi r(h + r) = 2\pi rh + 2\pi r^2$$

is to be minimized and

$$2\pi rh + 2\pi r^2 = \pi rh + \pi rh + 2\pi r^2,$$

it follows that

$$\frac{\pi rh + \pi rh + 2\pi r^2}{3} \geqslant \sqrt[3]{2\pi^3 r^4 h^2}$$

or

$$\pi rh + \pi rh + 2\pi r^2 \geqslant 3\sqrt[3]{2\pi^3 r^4 h^2}.$$

The left-hand side is minimal when the equality holds; that is, only if $\pi rh = \pi rh = 2\pi r^2$. It follows that the surface area is minimal when $\pi rh = 2\pi r^2$, or $h = 2r$.

The levels of topic differentiation in the preceding two examples should not be viewed as prescriptive but rather as suggestive of the range of possible levels of treatment by which all students could be provided an opportunity to learn important mathematical ideas.

Additional examples of how topics in the core curriculum may be treated at differing levels of abstraction can be found in the discussion sections of the standards on algebra, trigonometry, and probability. When interpreting the examples of content differentiation, it is important to understand that all students are not expected to progress from the informal work usually associated with levels 1 or 2 to the advanced work associated with levels 4 or 5. Some students may begin and complete the consideration of a topic or problem at level 1 and then revisit the topic or problem later in the curriculum at a higher level, consistent with their growing understanding of mathematics. Others may begin the consideration of a topic at level 1 and then progress to work at a higher level within the course of a single unit. Depending on their interests and performance, a subset of this group of students will consider the topic in its full and rich detail at an appropriate point in the curriculum.

Summary

The preceding sections have identified the salient features of a proposed 9–12 curriculum that accepts as its starting point the position that there is a common core of mathematical ideas that all students should have an opportunity to learn. The following standards identify and describe the content that should be included in such a curriculum. When these standards are reviewed and interpreted, it is important to remember that they should be viewed in the context of the mathematics content and approaches to that content specified in the curriculum standards for grades K–8. It is equally important to understand that the standards have been differentiated by strands for purposes of emphasis and discussion. Although the labels of several of the strands are the same as the titles of existing courses, the reader should not interpret those standards as course descriptions, but rather as topical strands that could be integrated across courses. A compelling rationale for, and outlines of, integrated mathematics programs can be found in Hirsch (1985).

STANDARD 1: MATHEMATICS AS PROBLEM SOLVING

In grades 9–12, the mathematics curriculum should include the refinement and extension of methods of mathematical problem solving so that all students can—

- ***use, with increasing confidence, problem-solving approaches to investigate and understand mathematical content;***
- ***apply integrated mathematical problem-solving strategies to solve problems from within and outside mathematics;***
- ***recognize and formulate problems from situations within and outside mathematics;***
- ***apply the process of mathematical modeling to real-world problem situations.***

Focus

Mathematical problem solving, in its broadest sense, is nearly synonymous with doing mathematics. Thus, whereas it is useful to differentiate among conceptual, procedural, and problem-solving goals for students in the early stages of mathematical learning, these distinctions should begin to blur as students mature mathematically. In grades 9–12, the problem-solving strategies learned in earlier grades should have become increasingly internalized and integrated to form a broad basis for the student's approach to doing mathematics, regardless of the topic at hand. From this perspective, problem solving is much more than applying specific techniques to the solution of classes of word problems. It is a process by which the fabric of mathematics as identified in later standards is both constructed and reinforced.

Discussion

One consequence of students' increasing mathematical sophistication is that problem situations, which for younger students necessarily arise from the real world, now often spring from within mathematics itself. Thus, mathematical problem solving serves not only to answer questions raised in everyday life, in the physical and social sciences, and in such professions as business and engineering but also to further extend and connect mathematical theory itself. A student who proves a theorem in order to extend knowledge in an axiomatic system and one who solves an application involving an optimal production and marketing decision have each engaged in varying levels of mathematical problem solving.

Problems and applications should be used to introduce new mathematical content, to help students develop both understanding of concepts and facility with procedures, and to apply and review processes they have already learned. For example, a situation such as finding the maximum height of the path of a projectile might be posed for which students have no readily available solution techniques. The learning process would require them to analyze the situation in light of their existing knowledge, develop appropriate mathematical techniques, and subsequently apply those techniques to solve the problem. "Looking back" over the problem situation and the whole problem-solving process also can provide a springboard from which even more efficient solution methods or problem extensions can be developed in ways that mathematically enrich the students' experience. This scenario may take place over a few days or even a few weeks; it often may be appropriate for students to work cooperatively in groups. It is the intent of this standard that this process be repeated across the curriculum on a regular and sustaining basis and that it entail appropriate student use of calculator and computer technology.

Students in grades 9–12 should also have some experience recognizing and formulating their own problems, an activity that is at the heart of doing mathematics. For example, an exploration of the perimeters of various rectangles with area 24 cm^2 by means of models or drawings, with data as recorded in table 1.1, could lead to student recognition and formulation of such problems as the following: Is there a rectangle of minimum perimeter with the specified area? What are its dimensions?

TABLE 1.1
Rectangle Data

Area	Length	Width	Perimeter
24 cm^2	1 cm	24 cm	50 cm
24 cm^2	2 cm	12 cm	28 cm
24 cm^2	3 cm	8 cm	22 cm
24 cm^2	4 cm	6 cm	20 cm
24 cm^2	6 cm	4 cm	20 cm
24 cm^2	8 cm	3 cm	22 cm

Instructional settings that encourage investigation, cooperation, and communication foster problem posing as well as problem solving. In addition, all students can profit from discussions of specific problem-posing techniques. Forming the "dual" of the problem above leads to the question, Is there a rectangle of *maximum area* with a *specified perimeter*? Other useful techniques include relaxing conditions in, or generalizing from, problem situations and considering the converse of mathematical statements.

Another important component of mathematical thinking is the process of mathematical modeling as illustrated in figure 1.1.

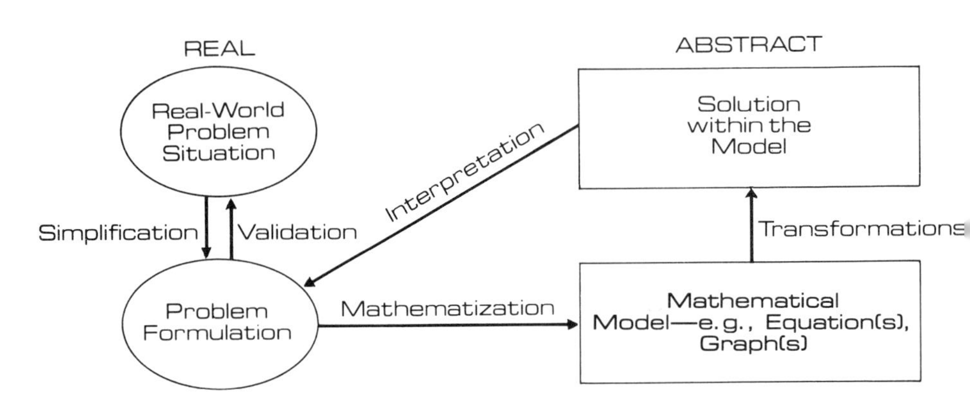

***Fig. 1.1.* Mathematical modeling**

The various stages in building and using a mathematical model are exemplified in the following solution of a famous problem first posed in the ninth century and finally solved 800 years later in 1654 by the famous French mathematicians Pierre Fermat and Blaise Pascal. Observe that the problem comes from probability, but the mathematical model is geometric. (*Note:* We have modified the content but not the nature of the original problem setting.)

♦ ♦ ♦ ♦ ♦ ♦ ♦ ♦

Real-world problem situation. **In a two-player game, one point is awarded at each toss of a fair coin. The player who first attains *n* points wins a pizza. Players A and B commence play; however, the game is interrupted at a point at which A and B have unequal scores. How should the pizza be divided fairly? (The intuitive division, that A should receive an amount in proportion to A's score divided by the sum of A's score and B's score, has been determined to be inequitable.)**

Problem formulation. Consider the situation with the following data: The winning score is $n = 10$; when the interruption occurs, the score is A:8 and B:7. The pizza will be divided in proportion to each player's probability of winning the game.

Mathematical model. See figure 1.2. At each turn, P(A wins a point) = P(B wins a point) = 1/2. A's share = P(A wins 10 points) × area of pizza; B's share = total pizza − A's share. Let a square region represent the original game state with the score A:8 to B:7 as indicated. At each turn, the square or interior rectangles are halved to represent $P = 1/2$ for winning (or losing) a point. Thus, in this model the resulting fraction of the original area also represents the probability of reaching that game state.

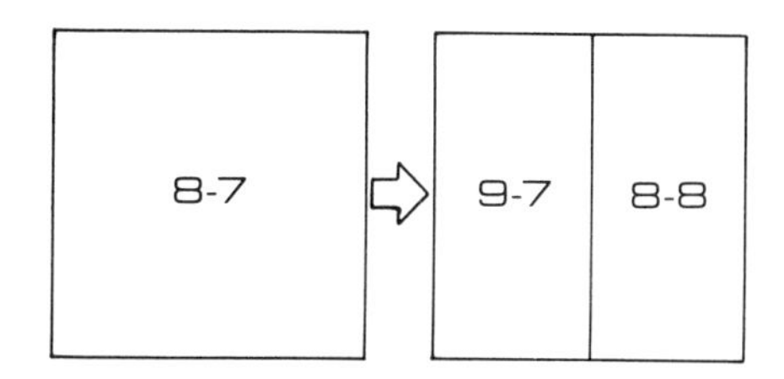

Fig. 1.2. A geometric probability model

Solution within the model. See figure 1.3.

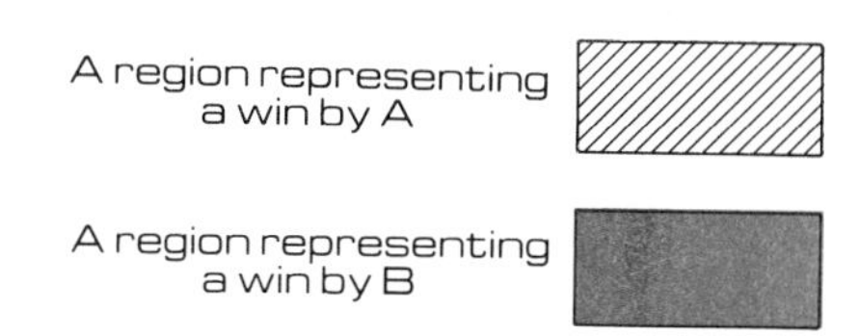

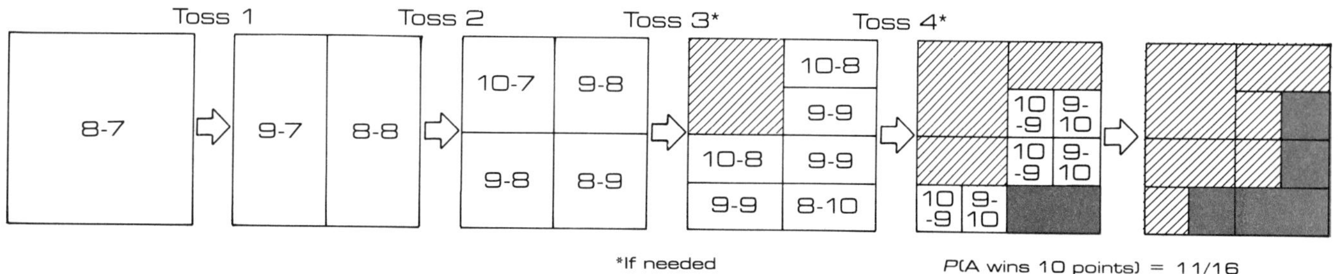

P(A wins 10 points) = 11/16

Fig. 1.3. Solution using the model

Interpretation of solution in original problem formulation

A's share = 11/16 of pizza
B's share = 5/16 of pizza

Validation in original real-world problem situation. Empirical evidence gained from actually playing out the game many times or, more easily, from computer simulation (using random numbers to represent coin tosses) confirms this solution. Simulation techniques are further illustrated in the standards on mathematical connections and probability.

The importance of problem solving to all education cannot be overestimated. To serve this goal effectively, the mathematics curriculum must provide many opportunities for all students to meet problems that interest and challenge them and that, with appropriate effort, they can solve.

STANDARD 2: MATHEMATICS AS COMMUNICATION

In grades 9–12, the mathematics curriculum should include the continued development of language and symbolism to communicate mathematical ideas so that all students can—

- ***reflect upon and clarify their thinking about mathematical ideas and relationships;***
- ***formulate mathematical definitions and express generalizations discovered through investigations;***
- ***express mathematical ideas orally and in writing;***
- ***read written presentations of mathematics with understanding;***
- ***ask clarifying and extending questions related to mathematics they have read or heard about;***
- ***appreciate the economy, power, and elegance of mathematical notation and its role in the development of mathematical ideas.***

Focus

All students need extensive experience listening to, reading about, writing about, speaking about, reflecting on, and demonstrating mathematical ideas. Active student participation in learning through individual and small-group explorations provides multiple opportunities for discussion, questioning, listening, and summarizing. Using such techniques, teachers can direct instruction away from a focus on the recall of terminology and routine manipulation of symbols and procedures toward a deeper conceptual understanding of mathematics. It is not enough for students to write the answer to an exercise or even to "show all their steps." It is equally important that students be able to describe how they reached an answer or the difficulties they encountered while trying to solve a problem. Continually encouraging students to clarify, paraphrase, or elaborate is one means by which teachers can acknowledge the merit of students' ideas and the importance of their own language in explaining their thinking. Providing opportunities for discussions about issues, people, and the cultural implications of mathematics reinforces student understanding of the connection between mathematics and our society.

In grades 9–12, methods of mathematical communication become more formal and symbolic. Facility with mathematical language and notation enables students to more easily form multiple representations of ideas, express relationships within and among representation systems, and formulate generalizations. In fact, facility with the language of mathematics is an integral part of thinking mathematically, solving problems, and reflecting on one's own mathematical experiences.

Discussion

Although K–8 students will have many experiences with informal use of language and the construction of arguments, at the high school level these experiences are extended to the use of specialized symbolism associated with the various representation systems of mathematics. However, the introduction and use of technical symbolism should evolve as a natural extension and refinement of the students' own language. Students in grades 9–12 should build on their informal reasoning experiences in previous grades to write convincing arguments that validate their own generalizations. College-intending students should be able to extend those arguments to deductive proofs in which underlying inference schemes are made explicit.

The following example illustrates how the views of mathematics as problem solving, communication, and reasoning are inextricably connected.

Nine robots (fig. 2.1) are to perform various tasks at fixed positions along an assembly line. Each must obtain parts from a single supply bin to be located at some point along the line. Students are asked to investigate where the bin should be located so that the total distance traveled by all the robots is minimal.

Fig. 2.1. Assembly line with nine robots

The investigation should include an opportunity for class discussion of possible reasons for attempting to minimize this distance.

Students can address this problem by first experimenting with simpler cases, as in figure 2.2 They will determine that for $n = 2$, any point on $\overline{R_1R_2}$ (or at a more sophisticated level, any $P \in [R_1, R_2]$) will work. This conclusion can be expected to emerge only after extensive argument, since the "natural" point to consider is the midpoint. For $n = 3$, R_2 is the solution. For $n = 4$, any point on $\overline{R_2R_3}$ will give the minimum distance. (Students may reason that for R_1 and R_4, any point in $[R_1, R_4]$ will work. Similarly, for R_2 and R_3, any point in $[R_2, R_3]$ will work. Thus, the solution is in any point in $[R_1, R_4] \cap [R_2, R_3] = [R_2, R_3]$.) For $n = 5$, similar reasoning yields R_3 as the optimal point, and so on. It follows that the solution of the original problem is to locate the bin at the position corresponding to R_5.

R_1 R_2 | R_1 R_2 R_3 | R_1 R_2 R_3 R_4 | R_1 R_2 R_3 R_4 R_5

Fig. 2.2. Simpler cases

All students should be encouraged to generalize their solutions to the case of n robots. The language and notation used by students will vary with their mathematical sophistication. Although all students should be expected to express their generalizations accurately (e.g., "at the middlemost robot's position or between the two middlemost robots"), college-intending students should additionally be able to use symbolic notation:

If n is even, then any $P \in [R_{n/2}, R_{n/2+1}]$ is optimal.
If n is odd, then $R_{(n+1)/2}$ is the optimal point.

Contextual situations and student experiences similar to these will serve to enhance students' appreciation of the value of mathematical activity and instill confidence in their ability to make sense of new problem situations.

In addition to the mathematical symbols related to concepts and operations developed in grades K–8, students in grades 9–12 need to use a variety of new symbols related to arrays, functions, and probability. This expanded symbol system extends and refines a student's ability to express quantitative ideas concisely. For example, the jeans-supply matrix in figure 2.3 provides an economical and well-ordered way of representing the size-by-brand information for a particular jeans department.

	Guess	Klein	Lee	Levi	Polo
28"	3	5	2	1	6
30"	4	4	1	0	7
32"	0	3	2	2	6
34"	1	6	4	3	4
36"	4	2	2	3	1

Fig. 2.3. Jeans-supply matrix

College-intending students would be expected to use more sophisticated notation associated with functions (including transformations), iterative algorithms, matrices, complex numbers, series, and limits in preparation for their continued study of mathematics.

In grades 9–12, student learning of mathematics becomes increasingly self-directed and dependent on textual materials. Appropriate symbolism and vocabulary should be used in all material presented *to* students, with the clear expectation of the appropriate use of such symbolism and notation *by* students. It cannot be assumed that even students who are skilled readers can read mathematical exposition effectively. All students will need specific instruction on how to read mathematical textbooks with understanding and how to use textbooks as valuable resources. Assignments that require students to read mathematics and respond both orally and in writing to questions based on their reading should be an integral part of the 9–12 mathematics program.

Techniques used to teach writing can be useful in teaching mathematical communication. The view of writing as a process emphasizes brainstorming, clarifying, and revising; this view can readily be applied to solving a mathematical problem. The simple exercise of writing an explanation of how a problem was solved not only helps clarify a student's thinking but also may provide other students fresh insights gained from viewing the problem from a new perspective.

Students could be encouraged to keep journals describing their mathematical experiences, including reflections on their problem-solving thought processes. Journal writing also can help students clarify feelings about mathematics or about a particular experience or activity in a mathematics classroom. These activities can foster students' positive attitudes about mathematics, particularly if the journal entries are accompanied by discussions about any negative feelings and ways to deal with unpleasant experiences.

Technology is yet another avenue for mathematical communication, both in transmitting and receiving information. Calculators and computers require students to use and understand accurate, concise language. To use a calculator, students must not only understand the underlying mathematics (e.g., the order of operations or the meaning of the fraction line) but also apply the specific syntax for the type of calculator being used. Using a computer language to implement a mathematical procedure requires translating the language of mathematics into the language of programming and then applying the syntax of the particular programming language. Interpreting the output of a computer program or a calculator display requires students to recognize equivalent forms of representation and to judge the reasonableness of results. Interpreting computer and calculator graphic displays additionally requires careful attention to the scales on the axes and an understanding of the effects of scaling on the characteristics of a graph.

Students whose primary language is not the language of instruction have unique needs. Specially designed activities and teaching strategies (developed and implemented with the assistance of language specialists) should be incorporated into the high school mathematics program so that all students have the opportunity to develop their mathematical potential regardless of a lack of proficiency in the language of instruction.

STANDARD 3: MATHEMATICS AS REASONING

In grades 9–12, the mathematics curriculum should include numerous and varied experiences that reinforce and extend logical reasoning skills so that all students can—

- ***make and test conjectures;***
- ***formulate counterexamples;***
- ***follow logical arguments;***
- ***judge the validity of arguments;***
- ***construct simple valid arguments;***

and so that, in addition, college-intending students can—

- ***construct proofs for mathematical assertions, including indirect proofs and proofs by mathematical induction.***

Focus

Inductive and deductive reasoning are required individually and in concert in all areas of mathematics. A mathematician or a student who is doing mathematics often makes a conjecture by generalizing from a pattern of observations made in particular cases (inductive reasoning) and then tests the conjecture by constructing either a logical verification or a counterexample (deductive reasoning). It is a goal of this standard that all students experience these activities so that they come to appreciate the role of both forms of reasoning in mathematics and in situations outside mathematics. Furthermore, all students, especially the college intending, should learn that deductive reasoning is the method by which the validity of a mathematical assertion is finally established.

A second goal of this standard is to expand the role of reasoning, now addressed primarily in geometry, so that it is emphasized in all mathematics courses for all students. In addition, this standard proposes that college-intending students should learn the more formal methods of proof required for college-level mathematics.

A third goal, also a departure from the existing curriculum for college-intending students, is to give increased attention to proof by mathematical induction, the most prominent proof technique in discrete mathematics. (The term *mathematical induction* refers to a formal technique used to prove statements defined for subsets of the integers. It should not be confused with inductive reasoning.)

Discussion

In grades 5–8, students will experience inductive reasoning and the evaluation and construction of simple deductive arguments in a variety of problem-solving settings. In grades 9–12, as the depth and complexity of content is increased, this emphasis on the interplay between conjecturing and inductive reasoning and the importance of deductive verification should be maintained.

All students, for example, should examine numerical patterns that result from algebraic manipulation, make conjectures about general algebraic properties based on their observations, and verify their conjectures with numerical substitutions. The properties of logarithms provide a context to illustrate our meaning. Using their knowledge of exponents and the fact that $\log ab = \log a + \log b$, students could explore cases such as the following:

$$\log 5^2 = \log (5 \cdot 5) = \log 5 + \log 5 = 2 \log 5$$

$$\log 5^3 = \log (5^2 \cdot 5) = \log 5^2 + \log 5 = 2 \log 5 + \log 5 = 3 \log 5$$

$$\log 5^4 = \log (5^3 \cdot 5) = \log 5^3 + \log 5 = 3 \log 5 + \log 5 = 4 \log 5$$

Students could be asked to examine the emerging pattern and generalize to $\log 5^n = n \log 5$ for each nonnegative integer n. Students could then use the $\boxed{y^x}$ and $\boxed{\log}$ keys on a calculator to confirm this generalization with several numerical values such as $n = 9, 14$, and 0.

As in the setting decribed above, algebraic processing itself often suggests generalizations. Some students might observe, for example, that their algebraic manipulations did not depend on the particular number 5, thus leading to the more general assertion, $\log a^n = n \log a$ for any $a > 0$.

Students can be introduced to the forms of deductive argument by examining everyday situations in which such forms arise naturally. Political claims and commercial advertisements are especially good sources for arguments that illustrate logical errors. For instance, a sign on the front of an ice-cream store declares, "If you want the best ice cream in the country, try Great Northern." The sign is meant to entice people into the Great Northern Ice Cream store, presumably to get the best ice cream in the country. However, the *converse* of the statement on the sign actually is needed to argue (by modus ponens) that Great Northern ice cream is the best in the country. This sign merely asserts that those who want the best ice cream in the country (i.e., almost anyone) should try Great Northern. It makes no claim about the quality of Great Northern ice cream.

Informal deductive arguments like that illustrated in the preceding example are applicable in mathematical settings. Students can begin to appreciate the power of deductive reasoning by providing simple valid arguments as justification for their solutions to specific problems and for algorithms constructed for various purposes. For example, all students should be able to compute the Euclidean distance between points with coordinates (2, 3) and (−4, 5) using the distance formula; additionally, all students should be able to provide a valid argument to justify why the computed distance is correct. This argument would be based on an appropriate figure, the Pythagorean theorem, and the knowledge of how to compute the distance between two points on a horizontal and on a vertical number line. However, it need not follow a particular format, and it may be presented orally or in writing in the student's own words. College-intending students, however, should be expected to derive and write a general proof of the distance formula. In contrast to the earlier argument, the proof would require arbitrary coordinates, as illustrated in figure 3.1. The proof itself would entail a careful sequence of steps with each step following logically from an assumed or previously proved statement and from previous steps. In addition, an argument should be made that the general formula applies when the points lie on the same horizontal or vertical line.

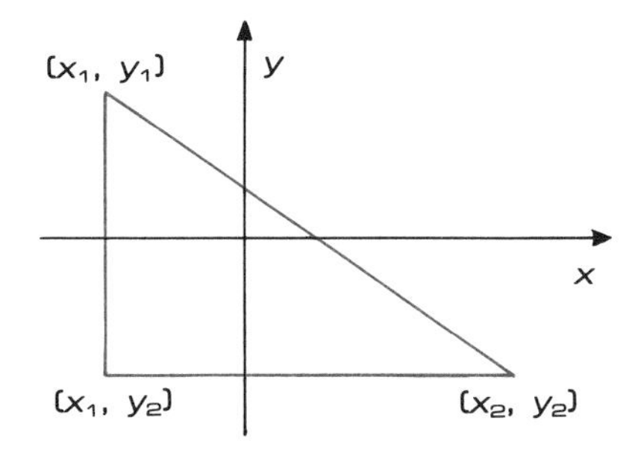

***Fig. 3.1.* Diagram with arbitrary coordinates**

Although reporting proofs in two-column form may be a useful teaching tool as students first learn to write proofs, eventually they should be ex-

◆ ◆ ◆ ◆ ◆ ◆ ◆ ◆

pected to write proofs in paragraph form. Only college-intending students should be expected to learn more specialized argument forms such as indirect proof and proof by mathematical induction. A proof of the conjecture that $\log a^n = n \log a$ for $a > 0$ and each nonnegative integer n would require students to use the principle of mathematical induction as follows:

1. For $n = 0$, $\log a^0 = \log 1 = 0 = 0 \log a$ for any $a > 0$.
   Thus, when $n = 0$, the equation is true.

2. Assume the equation is true for $n = k$ where k is any positive integer, that is, $\log a^k = k \log a$. Our goal is to show that the equation is true for the case $n = k + 1$. This is established by the following chain of reasoning:

   $$\log a^{k+1} = \log a^k a^1 = \log a^k + \log a = k \log a + \log a = (k + 1) \log a$$

   Thus, if the equation is true for $n = k$, then it is true for $n = k + 1$. It follows by the principle of mathematical induction that $\log a^n = n \log a$ for $a > 0$ and for every nonnegative integer n.

It is important that college-intending students have numerous and varied experiences with this proof technique in contexts beyond the familiar setting of series.

In grades 9–12, college-intending students will have their first experience of reasoning within an axiomatic system, the context that is so essential for work in mathematics. This higher-order thinking may not come easily, since the requirement to verify statements with a deductive proof by reasoning from axioms is unique to the discipline of mathematics; hence, it is a completely new way of thinking for high school students. Their previous experience both in and out of school has taught them to accept informal and empirical arguments as sufficient. Students should come to understand that although such arguments are useful, they do not constitute a proof.

It is also important that students recognize the difference between a statement that is verified by mathematical proof (i.e., a theorem) and one that is verified empirically using statistical arguments. Most mathematical theorems have a (sometimes hidden) quantifier, "for every." Thus, for example, the angle sum of *every* triangle in Euclidean geometry is 180 degrees, and $\sin^2 x + \cos^2 x = 1$ for *every* real number x. However, statements like "men are taller than women" or "women score higher than men on certain vocabulary tests" are true only in the sense that, in most empirial studies of these phenomena, the mean for a sample of men has differed from the mean for a sample of women at a level greater than some specified chance difference. In both cases, the men's and women's distributions actually overlap a great deal; that is, some women are taller than some men, and some men score higher on vocabulary tests than some women. It is not just incorrect logic to assume that statements arising from empirical and statistical arguments apply to every member of the groups in question; this error, when widely accepted in a society, contributes to inaccurate racial, ethnic, and gender sterotyping. A discussion of these reasoning issues can serve as a connection between mathematics and social studies classes in which the (usually negative) social consequences of such stereotyping are studied.

The potential for transfer between mathematical reasoning and the logic needed to resolve issues in everyday life can be enhanced by explicitly subjecting assertions about daily affairs to analysis in terms of the underlying principles of reasoning.

STANDARD 4: MATHEMATICAL CONNECTIONS

In grades 9–12, the mathematics curriculum should include investigation of the connections and interplay among various mathematical topics and their applications so that all students can—

- ***recognize equivalent representations of the same concept;***
- ***relate procedures in one representation to procedures in an equivalent representation;***
- ***use and value the connections among mathematical topics;***
- ***use and value the connections between mathematics and other disciplines.***

Focus

This standard emphasizes the importance of the connections among mathematical topics and those between mathematics and other disciplines, connections that are alluded to in many of the other standards. Two general types of connections are important: (1) modeling connections between problem situations that may arise in the real world or in disciplines other than mathematics and their mathematical representation(s); and (2) mathematical connections between two equivalent representations and between corresponding processes in each. These connections are illustrated in figure 4.1.

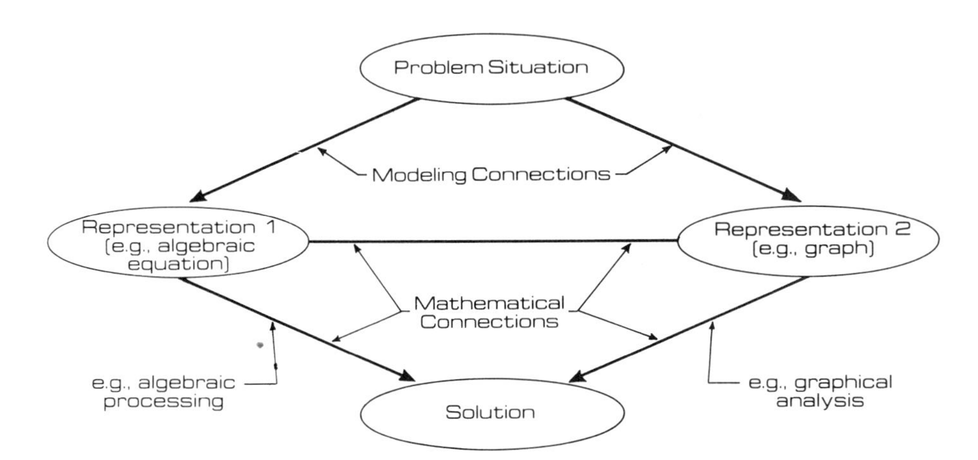

Fig. 4.1. Two general types of connections

Students who are able to apply and translate among different representations of the same problem situation or of the same mathematical concept will have at once a powerful, flexible set of tools for solving problems and a deeper appreciation of the consistency and beauty of mathematics.

It is important to recognize that the effectiveness of informal exploration depends on the extent to which it enables students to recognize crucial connections between informal activities and the mathematical ideas

♦ ♦ ♦ ♦ ♦ ♦ ♦ ♦

those activities are meant to convey. For example, exploratory geometric constructions help to develop understanding of formal geometric concepts only if students grasp the connections between straightedge and compass procedures and their formal geometric analogs.

Discussion

As in the earlier grades, teachers in grades 9–12 should introduce a new topic by exploring appropriate concrete representations in which students recognize that the exploratory activities embody the mathematical topic. This helps establish modeling connections, which can be further strengthened by an instructional approach that encourages multiple methods of solution for any given problem. Students should constantly be encouraged to look back at the solution process to consider other possible strategies. Such an approach helps establish mathematical connections as well by focusing students' attention on commonalities across different mathematical representations.

Students' understanding of the connections among mathematical ideas facilitates their ability to formulate and deductively verify conjectures across topics, an activity that becomes increasingly important in grades 9–12. In turn, these newly developed mathematical concepts and procedures can be applied to solve other problems arising from within mathematics and from other disciplines. The pervasiveness of the connections between mathematics and other disciplines is only hinted at by the following brief list of applications:

- *Art:* the use of symmetry, perspective, spatial representations, and patterns (including fractals) to create original artistic works
- *Biology:* the use of scaling to identify limiting factors on the growth of various organisms
- *Business:* the optimization of a communication network
- *Industrial arts:* the use of mathematics-based computer-aided design in producing scale drawings or models of three-dimensional objects such as houses
- *Medicine:* modeling an inoculation plan to eliminate an infectious disease
- *Physics:* the use of vectors to address problems involving forces
- *Social science:* the use of statistical techniques in predicting and analyzing election results

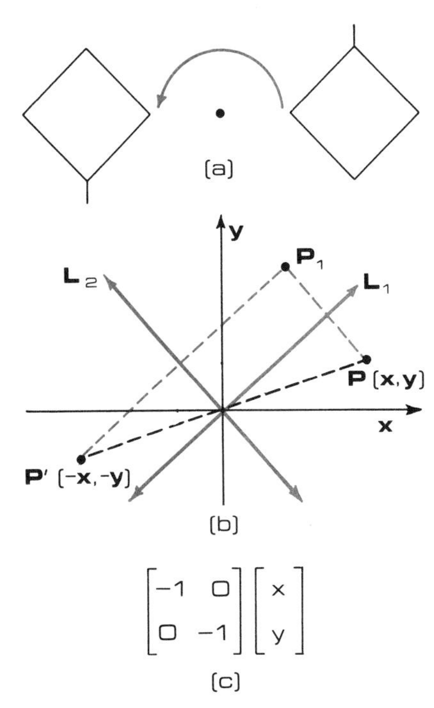

Fig. 4.2. Multiple representations of a half-turn

As a means of emphasizing the connections between mathematical ideas, new concepts should be introduced, where possible, as extensions of familiar mathematics. For example, the definitions of the trigonometric functions in terms of the coordinates of a point on the terminal side of an angle in standard position should be viewed as a natural extension of the trigonometric ratios defined for right triangles. In addition, the relationships among different representations of the same concept should be explored. Students, for example, may discover that the motion of a half-turn in geometry (fig. 4.2a) can be represented as a composition of reflections across two perpendicular lines or as the function $H[(x, y)] = (-x, -y)$ (fig. 4.2b), and as a matrix product (fig. 4.2c).

The connections between algebra and geometry are among the most important in high school mathematics. For example, finding $|a - b|$ corresponds to determining the distance between coordinates a and b on a number line; finding a zero of a function in algebra corresponds to determining an x-intercept of the graph of the function; finding a solution to a system of equations in algebra corresponds to determining the coordinates of the point(s) of intersection of the graphs of the equations; determining a local maximum value of a function in algebra corresponds to determining a local high point on the graph of the function; expressing a function in the form $y = af(bx + c) + d$ identifies the transformations that relate its graph to that of the simpler function $y = f(x)$; and approximating the limit of a function at a point of discontinuity corresponds to investigating the behavior of the graph of the function near that point. Computer-graphing technology now makes it possible to exploit these connections; many problems traditionally solved using algebra can now be solved efficiently and in more general cases using the geometric representation and computer-graphing techniques. This approach removes algebraic manipulative skill as a prerequisite, thereby allowing all students to address interesting problems and explore important mathematical ideas.

As an illustration of the power of mathematical connections, consider the problem of estimating the area of a region under a curve.

Approximate the area of the region under the curve $y = 2^x$, above the x-axis, and between the lines $x = 1$ and $x = 3$.

How could this area be determined? One way all students could estimate the area is by determining the area of trapezoid *ABCD* (fig. 4.3).

***Fig. 4.3.* Trapezoidal estimate of the area under a curve**

Students could then subdivide the interval [1,3] into two, three, . . . subintervals and use similar geometric reasoning to obtain better approximations of the area by using two trapezoids, three trapezoids, and so forth. Of course, a calculator or computer would be used to do the computations. This development would lead to a natural discussion of limits by all students (table 4.1). College-intending students could further extend this development to foreshadow the study of integral calculus.

TABLE 4.1
Refined Estimates of the Area under the Curve

No. of Trapezoids	Area Estimate
1	10
2	9
3	8.810
4	8.743
5	8.712
10	8.670

(*Note:* The actual area is $\frac{6}{ln2}$ or approximately 8.656 17.)

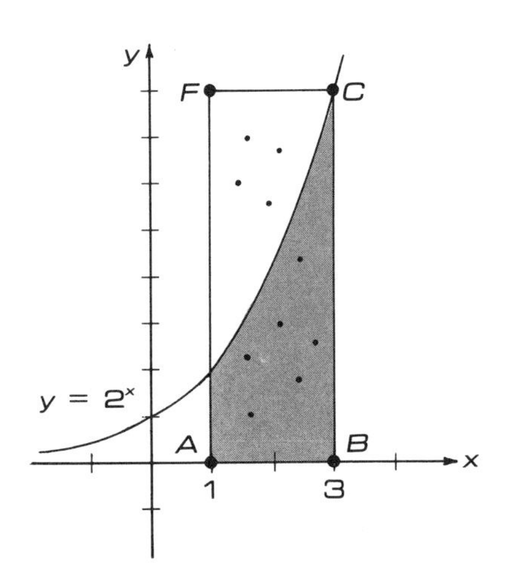

***Fig. 4.4.* Probabilistic model for estimating area under the curve**

A completely different approach is available to all students by using computer simulation to generate random points contained within the rectangle *ABCF* (fig. 4.4). This method is based on a probabilistic model that assumes that as the number of randomly generated points increases, the ratio on the right-hand side of the following equation will more closely approximate that on the left-hand side.

$$\frac{\text{Area under curve}}{\text{Area of } ABCF} \doteq \frac{\text{Total no. of points under or on curve}}{\text{Total no. of points}}$$

Table 4.2 reports estimates of the area under the curve for increasing numbers of computer-generated points.

This example highlights only some of the many connections among algebra, geometry, probability, and fundamental ideas of calculus.

◆ ◆ ◆ ◆ ◆ ◆ ◆ ◆

TABLE 4.2
Probabilistic Estimates of Area under the Curve

No. of Pts.	No. Pts. under Curve	Area Estimate
10	6	9.6
100	54	8.64
1 000	548	8.676 8
10 000	5 429	8.686 4

College-intending students should understand and be able to use connections among mathematical topics, including the following:

- Relations and functions
- Systems of equations and matrices
- Function equations expressed in standardized form and geometric transformations
- Complex numbers represented as $a + bi$ or $r(\cos \theta + i \sin \theta)$ and as ordered pairs (a,b) or (r,θ) in the complex plane
- Right-triangle ratios, trigonometric functions, and circular functions
- Circular functions and series
- Rectangular coordinates and polar coordinates
- Algorithms and their computer implementation
- Explicit and parametric representation of equations
- A function and its inverse, such as the logarithmic and exponential functions
- Statistical procedures and their requisite probability concepts
- The conversion of data to z-scores and geometric transformations
- Finite graphs and matrices
- Recursive and closed-form definitions of the same sequence

Instruction that focuses on networks of mathematical ideas rather than solely on the nodes of the networks in isolation will serve to instill in students an understanding of, and appreciation for, both the power and the beauty of mathematics. Developing mathematics as an integrated whole also serves to increase the potential for retention and transfer of mathematical ideas. Connecting mathematics with other disciplines and with daily affairs underscores the utility of the subject.

STANDARD 5: ALGEBRA

In grades 9–12, the mathematics curriculum should include the continued study of algebraic concepts and methods so that all students can—

- ***represent situations that involve variable quantities with expressions, equations, inequalities, and matrices;***
- ***use tables and graphs as tools to interpret expressions, equations, and inequalities;***
- ***operate on expressions and matrices, and solve equations and inequalities;***
- ***appreciate the power of mathematical abstraction and symbolism;***

and so that, in addition, college-intending students can—

- ***use matrices to solve linear systems;***
- ***demonstrate technical facility with algebraic transformations, including techniques based on the theory of equations.***

Focus

Algebra is the language through which most of mathematics is communicated. It also provides a means of operating with concepts at an abstract level and then applying them, a process that often fosters generalizations and insights beyond the original context.

Aspects of this standard represent extensions of algebraic concepts developed first in grades 5–8. Whereas this earlier work was developed as a generalization of arithmetic, algebra in grades 9–12 will focus on its own logical framework and consistency. As a result, for example, algebraic symbols may represent objects rather than numbers, as in "$p + q$" representing the sum of two polynomials. This more sophisticated understanding of algebraic representation is a prerequisite to further formal work in virtually all mathematical subjects, including statistics, linear algebra, discrete mathematics, and calculus. Moreover, the increasing use of quantitative methods, both in the natural sciences and in such disciplines as economics, psychology, and sociology, have made algebraic processing an important tool for applying mathematics.

The proposed algebra curriculum will move away from a tight focus on manipulative facility to include a greater emphasis on conceptual understanding, on algebra as a means of representation, and on algebraic methods as a problem-solving tool. For the core program, this represents a trade-off in instructional time as well as in emphasis. For college-intending students who can expect to use their algebraic skills more often, an appropriate level of proficiency remains a goal. Even for these students, however, available and projected technology forces a rethinking of the level of skill expectations.

Discussion

Algebra as a means of representation is most readily seen in the translation of quantitative relations to equations or graphs. For example, to re-

♦ ♦ ♦ ♦ ♦ ♦ ♦ ♦

late auto speed to stopping distance, collected data could be organized as in table 5.1 and analyzed for patterns.

TABLE 5.1
Automobile Stopping Distance

Speed (in mph)	Reaction Distance (in ft.)	Braking Distance (in ft.)	Stopping Distance (in ft.)
10	10	5	15
20	20	20	40
30	30	45	75
40	40	80	120
50	50	125	175
60	60	180	240

From this particular set of data, students could deduce that if s represents speed, then the representations for the reaction, braking, and stopping distances are s, $s^2/20$, and $s^2/20 + s$, respectively. The equation $d = s^2/20 + s$ would provide a problem-solving tool for interpolating and extrapolating values not included in the original table of collected data. (Corresponding activities could be applied to the equations modeling the data in each of the second and third columns of the table.) Follow-up project work for students could include preparing an oral or written report comparing this algebraic model with the usual "rule of thumb" cited in driver's education classes or researching data on braking distances for autos equipped with disc brakes or with antilock braking systems and then developing corresponding equations relating speed and stopping distance.

Situations in which there is a great amount of numerical data to be recorded and manipulated, such as with factory (store) inventories, production (sales) figures, and shipments, often are represented by matrices. For example, if I represents the initial jeans-inventory matrix (fig. 2.3) on page 141, P the sales matrix, and S the shipment matrix on a given day, then $I - P + S$ is the matrix representation of the inventory at the end of the business day. Matrix representations of data permit easy processing by computers and thus have become important representation tools in algebra.

Changes in emphases require more than simple adjustments in the amount of time to be devoted to individual topics; they also will mean changes in emphases *within* topics. For example, although students should spend less time simplifying radicals and manipulating rational exponents, they should devote more time to exploring examples of exponential growth and decay that can be modeled using algebra. Similarly, students should spend less time plotting curves point by point, but more time interpreting graphs, exploring the properties of graphs, and determining how these properties relate to the forms of the corresponding equations (e.g., the relationship between the graphs of $y = |x|$ and $y = |x - 5|$). Of course, students should continue to plot critical points to check the reasonableness of graphs.

Computing technology enables schools to provide a richer set of algebra experiences for all students. Polynomial equations, which are very useful for describing relations among variables in a vast array of real-world situations, need no longer be a topic reserved for precalculus students. To illustrate, consider the box-building activity (p. 80) described in the grades 5–8 standard on communication. This activity would be extended in grades 9–12 to boxes similarly produced by cutting squares from the corners of *rectangular* sheets. If the dimensions of a sheet are 25 inches by 40 inches and the length of the side of the squares is x inches, then the cubic equation $V = x(25 - 2x)(40 - 2x)$ describes the relationship between the volume and the height of the resulting box. To de-

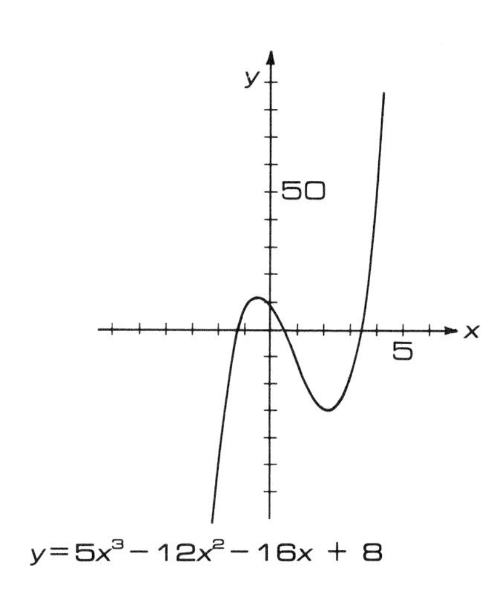

$y = 5x^3 - 12x^2 - 16x + 8$

Observe that the roots are between
−2 and −1
0 and 1
3 and 4.

***Fig. 5.1.* Bounds for roots**

termine a value x for which the volume was a specified number, say, $V = 1800$ cubic inches, would require solving the equation $x(25 - 2x)(40 - 2x) = 1800$ or equivalently the equation $x(25 - 2x)(40 - 2x) - 1800 = 0$. Similar equations frequently arise in the management sciences in the process of analyzing cost, revenue, and profit in the production and sale of goods. Problems of this sort lead naturally to the question, "How does one solve an equation like $ax^3 + bx^2 + cx + d = 0$?"

The following example illustrates how the treatment of polynomial equation-solving can be differentiated in both depth and the level of formalism so that all students in the core curriculum can experience success commensurate with their interests and proficiencies.

Find the roots of the equation $5x^3 - 12x^2 - 16x + 8 = 0$.

Level 1: Students would use either a table-building program or a graphing utility (fig. 5.1) to isolate the roots between pairs of consecutive integers.

They would then use a successive approximation method (either a refined search by altering the input values in the table-building program or guess and check with a calculator) to estimate the roots to the nearest tenth.

Level 2: Given the conceptual understanding and information gained in level 1 activities, students would use a built-in root-finding utility by simply entering the endpoints of the appropriate unit intervals.

Level 3: Students at this level would use a graphing zoom-in process to approximate the roots to the desired degree of accuracy, subject to machine precision. Figure 5.2 illustrates how this process is used to find the negative solution with error less than 0.001. (Note that if students use only an automated zoom-in feature, this level of mathematical activity corresponds to that at level 2. The use of a zoom-in feature that requires students to interpret the scales or a viewing rectangle in selecting the appropriate x- and y-intervals for the next nested viewing rectangle requires additional mathematical sophistication, appropriate to level 3.)

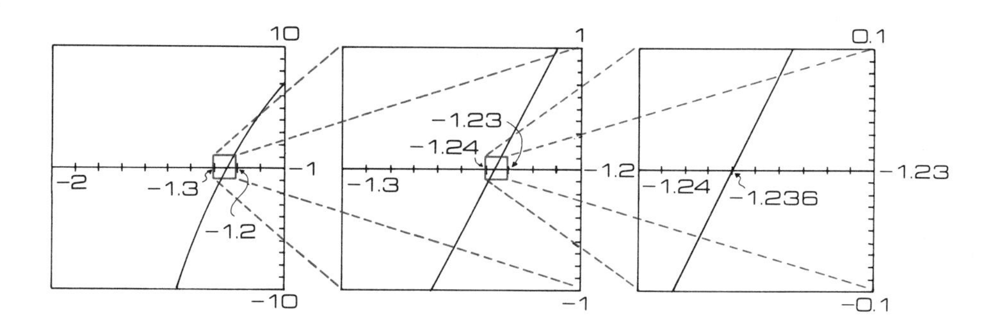

***Fig. 5.2.* Graphing zoom-in to approximate roots**

Level 4: After using a graphing utility as in level 1, students would be assigned a group project of constructing an algorithm for approximating roots, such as the following bisection algorithm. By considering several equations and investigating the relationship among equation values at, and on either side of, the midpoint of a unit interval under consideration, students can discover the pivotal idea underlying the algorithm.

♦ ♦ ♦ ♦ ♦ ♦ ♦ ♦

To estimate, to the nearest thousandth, the root of a polynomial equation known to be in an interval (x_1, x_2) whose endpoints are consecutive integers:

Input x_1, x_2
While $|x_1 - x_2| > .001$ Do
  $x_m \leftarrow \dfrac{x_1 + x_2}{2}$
  If $y_1 * y_m < 0$, then
    $x_2 \leftarrow x_m$
  Else $x_1 \leftarrow x_m$
Output x_m

(We recommend the use of notation similar to that above for expressing algorithms rather than the use of more familiar (and cumbersome) flowchart representations.)

Once an algorithm has been proposed to solve the problem, students would test the procedure by computer implementation.

Level 5: Experience with a graphing utility as in level 1 would lead to theoretical considerations regarding the number and nature of the roots. In particular, students would develop and use the rational root theorem to find the rational root(s). (In this example, 2/5 is the only rational root.) The development of the factor theorem would provide the basis for expressing this polynomial as a product of a linear and a quadratic factor, which in turn permits the other exact roots to be found by the quadratic formula. This development could be extended later to a discussion of complex roots and the fundamental theorem of algebra.

The reader should note that not only does the use of technology permit the study of polynomial equations to begin with problem situations, it also emphasizes powerful successive approximation and graphic methods that can easily be generalized to other types of equations. Moreover, the formal analysis of polynomial algebra is the culmination (level 5) of student activity, not the beginning.

Whereas all students should use matrices as tools for representation and problem solving, college-intending students also should experience formal study of matrix algebra and its applications to the solution of linear systems. Matrix methods for the solution of 2×2 and 3×3 systems are easily generalized to m equations in n variables. Computer implementation of such algorithms permits these students access to richer and more realistic problems. Further examples of the use of matrices by all students as tools for representation and problem solving are included in the elaboration for the standard on discrete mathematics.

STANDARD 6: FUNCTIONS

In grades 9–12, the mathematics curriculum should include the continued study of functions so that all students can—

- ***model real-world phenomena with a variety of functions;***
- ***represent and analyze relationships using tables, verbal rules, equations, and graphs;***
- ***translate among tabular, symbolic, and graphical representations of functions;***
- ***recognize that a variety of problem situations can be modeled by the same type of function;***
- ***analyze the effects of parameter changes on the graphs of functions;***

and so that, in addition, college-intending students can—

- ***understand operations on, and the general properties and behavior of, classes of functions.***

Focus

The concept of function is an important unifying idea in mathematics. Functions, which are special correspondences between the elements of two sets, are common throughout the curriculum. In arithmetic, functions appear as the usual operations on numbers, where a pair of numbers corresponds to a single number, such as the sum of the pair; in algebra, functions are relationships between variables that represent numbers; in geometry, functions relate sets of points to their images under motions such as flips, slides, and turns; and in probability, they relate events to their likelihoods. The function concept also is important because it is a mathematical representation of many input-output situations found in the real world, including those that recently have arisen as a result of technological advances. An obvious example is the $\boxed{\sqrt{x}}$ key on a calculator.

Discussion

To establish a strong conceptual foundation before the formal notation and language of functions are presented, students in grades 9–12 should continue the informal investigation of functions that they started in grades 5–8. Later, concepts such as domain and range can be formalized and $f(x)$ notation can be introduced, but care should be taken to treat these as natural extensions of the initial informal experiences. A function can be described by a written statement, by an algebraic formula, as a table of input-output values, or by a graph. Students need to work with each of these representations and to translate among them.

Since functional relationships are encountered so frequently, the study of functions should begin with a sampling of those that exist in the students' world. Students should have the opportunity to appreciate the pervasiveness of functions through activities such as describing real-world relationships that can be depicted by graphs, reading and interpreting graphs, and sketching graphs of data in which the value of one variable depends on the value of another. For example, students could be provided the graph in figure 6.1 and asked to determine what real-world phenomenon might be represented by the graph (e.g., temperature of an oven). Interpretive activities might include asking them to give an approximate value for A and to indicate what it represents; to determine what event is occurring during the time interval from B to C; or to explain why the graph oscillates for times greater than C.

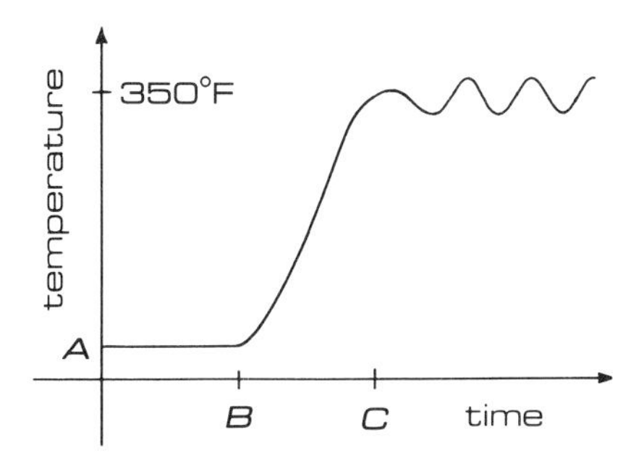

***Fig. 6.1.* Oven temperature as a function of time**

As a second example, consider the following student-prepared graph (fig. 6.2) and the richness of the setting for encouraging mathematical reasoning and communication. Students could be asked to explain a possible meaning of the vertical intercept (e.g., the cost of the band) or the break in the graph (e.g., the cost of a security officer required for more than 150 students), to determine how many tickets were sold before breaking even, to find the cost of the tickets sold prior to the dance and at the door, or to explain how the graph should be modified if it were discovered that the student treasurer failed to take into account the $50 cost of chaperones.

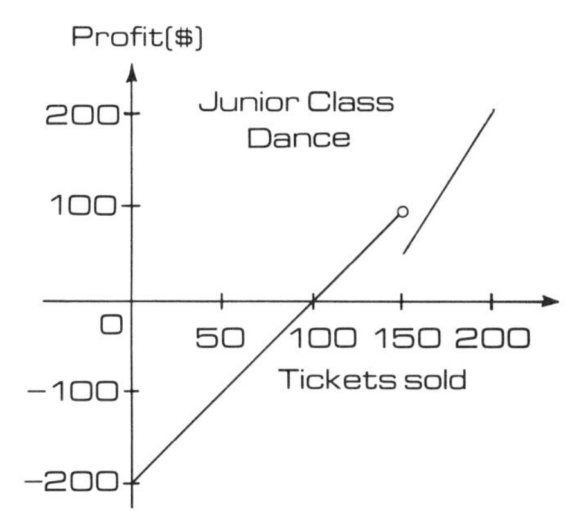

***Fig. 6.2.* Dance profit as a function of tickets sold**

Students are frequently given experiences graphing functions expressed in symbolic form. It is equally important, however, that they be given opportunities to translate from a graphical representation of a function to a symbolic form. For example, an analysis of the graph in figure 6.2 would lead to the following piecewise definition of the function:

$$\begin{aligned} P(t) &= 2t - 200 \text{ for } 0 \leqslant t < 150 \\ &= 3t - 400 \text{ for } 150 \leqslant t \leqslant 200 \end{aligned}$$

The power of functions to simplify complex situations and to predict outcomes can be demonstrated by observing a phenomenon involving an underlying functional relationship between two variables, gathering and plotting observational data, fitting a graph to the plotted points, using the graph to formulate the relationship between the variables, and then predicting outcomes for unobserved values of one of the variables. For example, students could record the number of swings during a given time period for pendulums of differing lengths, graph the relationship between the number of swings and the length of the pendulum, formulate this relationship, use it to predict the number of swings for pendulums of other lengths, and validate their predictions.

Computing technology provides tools, especially spreadsheets and graphing utilities, that make the study of function concepts and their applications accessible to all students in grades 9–12. This technology makes it possible for students to observe the behavior of many types of functions, including direct and inverse variation, general polynomial, radical, step, exponential, logarithmic, and sinusoidal. All students should use a graphing utility to investigate how the graph of $y = af(bx + c) + d$ is related to the graph of $y = f(x)$ for various changes of the parameters a, b, c, and d. The effects of these changes can best be expressed by geometric transformations, thereby providing another connection between algebra and geometry. This is illustrated for the function $f(x) = x^2$ by the sequence of graphs in figure 6.3.

College-intending students should develop an understanding of polynomial, rational, algebraic, and transcendental functions, and of those defined piecewise in terms of any of the above. Each of these types of functions should be used to model several problem situations so students can abstract the differences and commonalities in problem situations that are modeled by a given type of function.

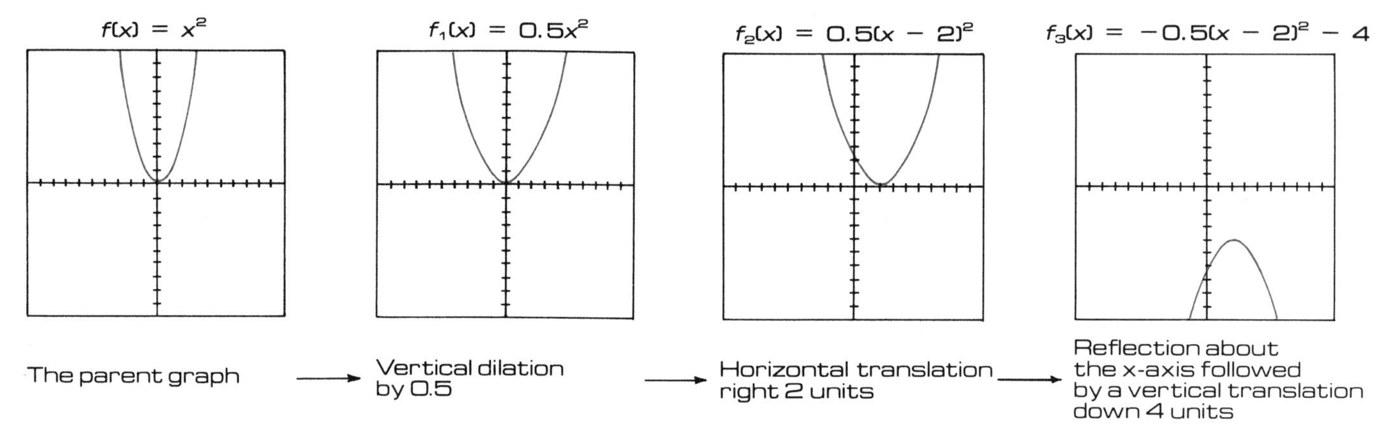

Fig. 6.3. Altering functions, transforming graphs

Computing technology has made recursively defined functions increasingly important. As a result, college-intending students should develop some facility with functions defined in this manner. For example, $n!$ can be viewed recursively as a function f defined on the nonnegative integers as follows:

$$f(0) = 0! = 1$$
$$f(n) = n! = nf(n - 1) = n(n - 1)! \text{ for } n \geq 1$$

Once they have acquired an understanding of various functions, college-intending students should apply techniques to fit curves, that is, functions, to data collected experimentally or supplied in tabular form. If the data are not linear, students should use finite-difference methods, statistical computer packages, and log-log or semi-log transformations to formulate and subsequently make predictions from the function of best fit.

College-intending students also should learn how to combine functions by addition and by composition, and properties of these operations should be analyzed and interpreted graphically. The concept of inverse function should be explored informally by all students as a process of undoing the effect of applying a given function. Furthermore, all students could engage in activities in which they discover and apply the fact that the reflection image of the graph of a one-to-one function across the line $y = x$ is the graph of its inverse. The formal notation and the precise definition of inverse function should be reserved for college-intending students.

Finally, college-intending students should use graphing utilities to investigate informally the surfaces generated by functions of two variables. Such investigations not only contribute to further development of important visualization skills but also foreshadow more advanced work with functions.

STANDARD 7: GEOMETRY FROM A SYNTHETIC PERSPECTIVE

In grades 9–12, the mathematics curriculum should include the continued study of the geometry of two and three dimensions so that all students can—

- ***interpret and draw three-dimensional objects;***
- ***represent problem situations with geometric models and apply properties of figures;***
- ***classify figures in terms of congruence and similarity and apply these relationships;***
- ***deduce properties of, and relationships between, figures from given assumptions;***

and so that, in addition, college-intending students can—

- ***develop an understanding of an axiomatic system through investigating and comparing various geometries.***

Focus

This component of the 9–12 geometry strand should provide experiences that deepen students' understanding of shapes and their properties, with an emphasis on their wide applicability in human activity. The curriculum should be infused with examples of how geometry is used in recreations (as in billiards or sailing); in practical tasks (as in purchasing paint for a room); in the sciences (as in the description and analysis of mineral crystals); and in the arts (as in perspective drawing).

High school geometry should build on the strong conceptual foundation students develop in the new K–8 programs. Students should have opportunities to visualize and work with three-dimensional figures in order to develop spatial skills fundamental to everyday life and to many careers. Physical models and other real-world objects should be used to provide a strong base for the development of students' geometric intuition so that they can draw on these experiences in their work with abstract ideas.

Discussion

Prior to the work of the ancient Greeks (e.g., Thales and Pythagoras), geometric ideas were tied directly to the solution of real-world problems. Hence, the subsequent abstraction and formalization of these ideas, which evolved into the subject of geometry as we know it, has always had many applications in the real world. More recently, fractal geometry, which originated in the mid-1970s with the pioneering work of Benoit Mandelbrot, has provided useful models for analyzing a wide variety of phenomena, from changes in coastlines to chaotic fluctuations in commodity prices. It is the intent of this standard that, whenever possible, real-world situations will provide a context for both introducing and applying geometric topics.

The construction trades use geometric ideas on a daily basis. When a builder is laying the foundation for a rectangular garage, opposite sides are measured to be the same length. However, it is also essential that

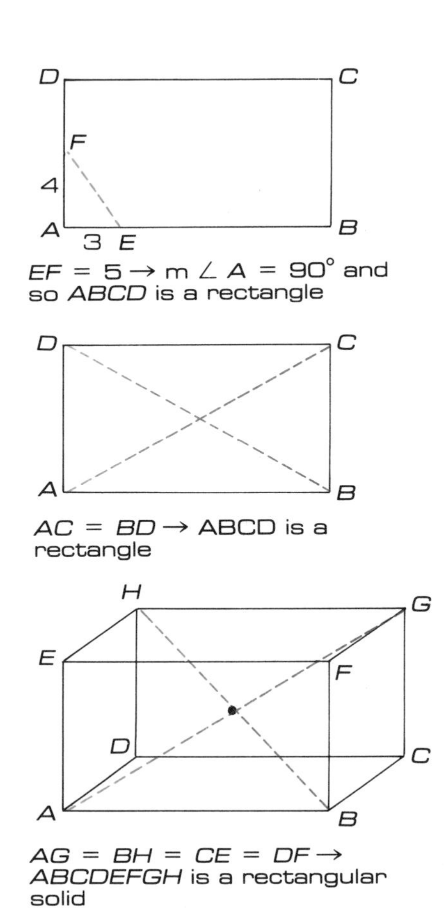

***Fig. 7.1.* Squaring up a rectangular foundation and building**

corner angles be right angles. Students could be encouraged to find ways to check whether this is so using only a tape measure. The two parallelograms on the top in figure 7.1 show two possible methods for solution; the second method, measuring the diagonals, also can be used to "square up" the frame of the garage if it is to be in the shape of a rectangular solid and it is known that opposite edges of the faces are the same length. If the diagonals are of equal length, the angles are right angles. Students may have the opportunity to apply these methods in actual construction in vocational classes or at home. Some students, especially the college intending, should be expected to cite or prove theorems to justify these methods.

Geometry also is used extensively in the sciences. In the biological sciences, for example, scaling (the geometric concept of similarity) is used to identify limiting factors on the growth of various organisms. Students who have learned some basic ideas of similarity, including the relationship between ratios of sides, areas, and volumes of similar figures, may find such practical, real-world applications of their geometry studies fascinating. Through discussion, students could agree that a very tall human being is roughly similar to one of more normal height, that weight is a function of volume, and that the ability to support weight is a function of the area of a cross-section of leg bones. Similarity concepts are then sufficient for the students to show that to enlarge a six-foot, 175-pound person by a scale factor of 2 would result in a twelve-foot giant, weighing 8×175 or 1400 pounds. The cross-sectional area of the giant's bones would be increased by a factor of 4. Thus, the pressure on the leg bones of a twelve-foot person who weighed 1400 pounds would be the same as that of a six-foot person who weighed 350 pounds. This explains why the size of human beings and other organisms is limited by their structural characteristics; the giant flies, spiders, and apes in science-fiction films could not exist in the real world. As an extension of this geometric method, students could read about and discuss practical size limits for inanimate objects such as trees, buildings, and mountains, which can be drawn from a knowledge of the strength of the supporting material such as wood, steel, and granite.

Geometry also provides an opportunity for students to experience the creative interplay between mathematics and art. For example, by repeated tracing of a regular polygon about a point so that the tracings coincide only along edges and do not overlap, and then by extending their tracings outward on the paper (plane), students can discover whether the polygon might be used to form a tiling (tessellation) of the plane. Out of this very informal experience arises a fundamental question: Which regular polygons can be used to tile a plane? This question, of course, leads to student investigation of how to determine the angle measure of a regular polygon, an opportunity for reasoning both deductively (from a knowledge of the angle-sum property of a triangle) and inductively. Once students have determined that only the equilateral triangle, square, and hexagon can be used singly to form a tiling pattern, project work for groups of students could include exploring (*a*) the number and nature of semiregular tilings using a combination of two or more regular polygons and where these patterns appear in their environment; (*b*) the existence of nonregular polygons that would serve as a fundamental tiling unit; or (*c*) the graphics work of M. C. Escher and the creation of Escher-type tessellations (fig. 7.2). This last activity is appealing to many high school students and provides an excellent setting for creative expression.

Instruction should focus increased attention on the analysis of three-dimensional figures. Such work is especially important to students who may pursue careers in art, architecture, drafting, and engineering. Appropriate use of three-dimensional representation and CAD (computer-assisted design) software is of particular value in such exercises. It is

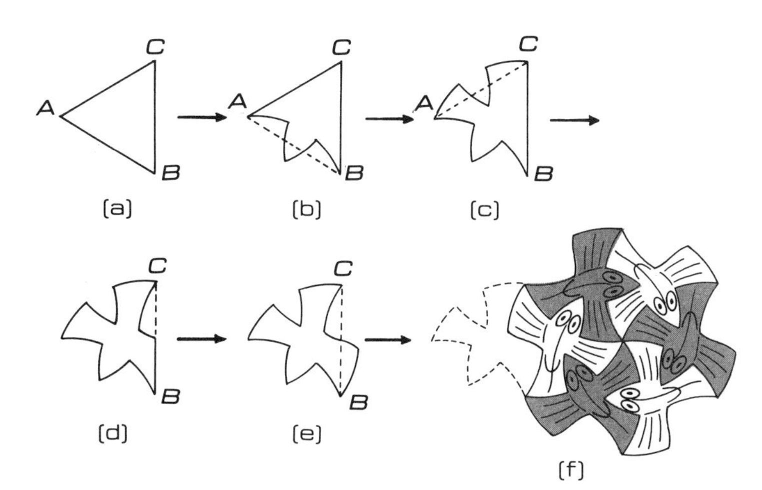

Fig. 7.2. Creating an Escher-type tessellation

also important to understand that visualization includes plane figures as well. For example, computer graphics software that allows students to create and manipulate shapes provides an exciting environment in which they can make conjectures and test their attempts at two-dimensional visualization. There are, of course, ample opportunities for visualization within standard activities that do not use a computer. In particular, exercises that require the student to represent given information by drawing a diagram provide an excellent setting for facilitating the reading of mathematics, as well as a special opportunity for problem translation.

Computer microworlds such as Logo turtle graphics and the topics of constructions and loci provide opportunities for a great deal of student involvement. In particular, the first two contexts serve as excellent vehicles for students to develop, compare, and apply algorithms.

Although the hypothetical deductive nature of geometry first developed by the Greeks should not be overlooked, this standard proposes that the organization of geometric facts from a deductive perspective should receive less emphasis, whereas the interplay between inductive and deductive experiences should be strengthened. For example, students should first use an interactive computer software package that allows experimentation with figures and relations to observe across several trials that the length of the median to the hypotenuse of any right triangle appears to be equal to the lengths of the segments it cuts off on the hypotenuse. In the second phase, they would provide a deductive argument verifying their discovery.

Both inductive and deductive reasoning are required as students begin to develop short sequences of theorems. For example, students could be provided the necessary definitions and a set of postulates consisting of three triangle-congruence statements (SSS, SAS, ASA) and a parallel-line statement (if two parallel lines are cut by a transversal, the alternate interior angles are congruent). After class discussion has established a definition of a parallelogram, students could be assigned the task of formulating and then proving or disproving their own conjectures about properties of parallelograms. Following agreement on definitions of a rectangle, square, and rhombus, students could discover and verify properties of these figures using the deduced properties of parallelograms. Students should perform these exercises without reference to similar proofs in a textbook. Figure 7.3 shows a sequence of theorems that could be deduced from these postulates.

Other topics amenable to organization by local axiomatics include theorems related to parallelism in a plane (in space), similarity of polygons, right-triangle relationships including the Pythagorean theorem, and areas of polygons.

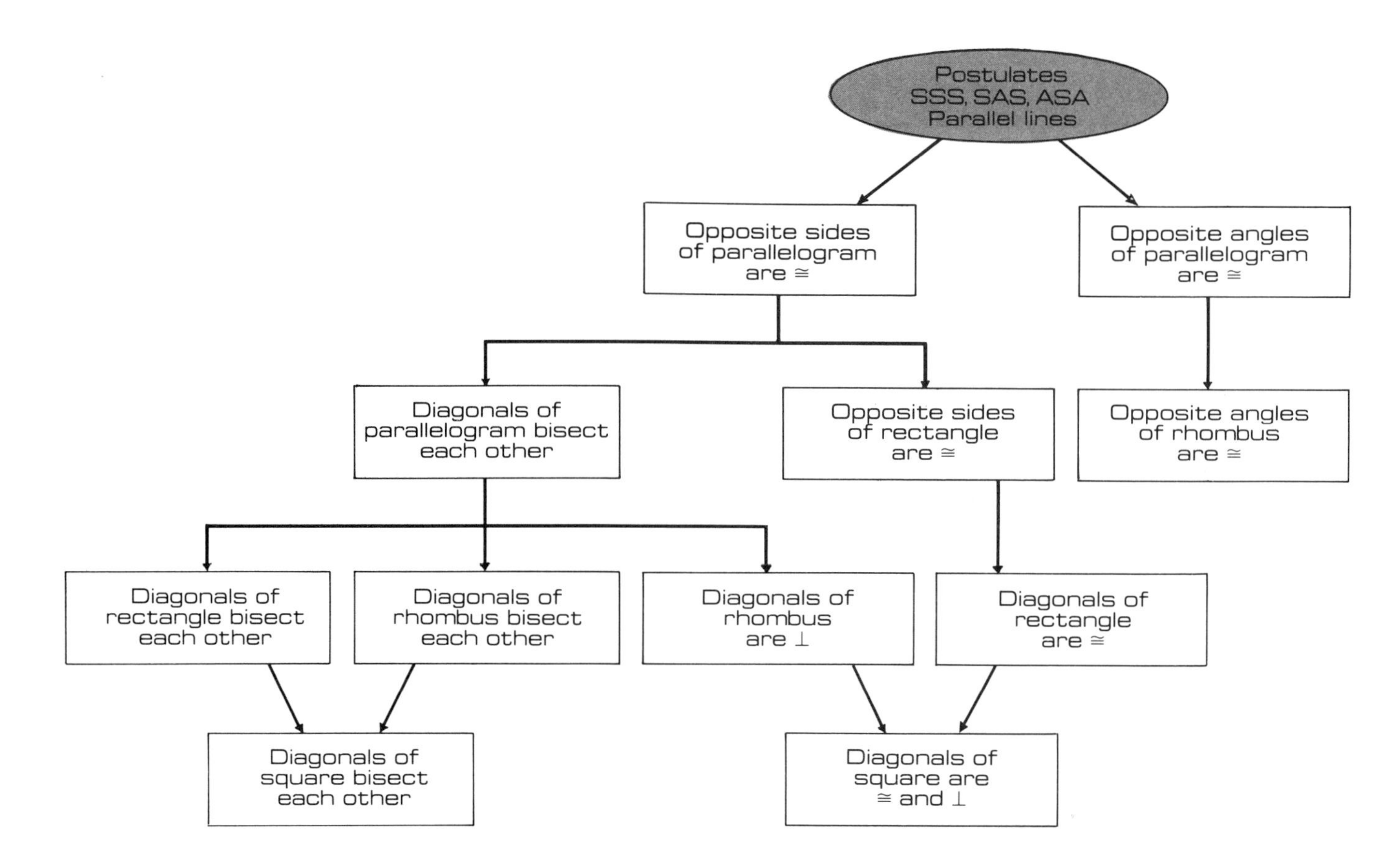

***Fig. 7.3.* An example of local axiomatics**

Logical reasoning also is conveyed through carefully designed numerical exercises. Consider, for example, the following exercise (for which the diagram in fig. 7.4 would *not* be provided):

In right triangle *ABC* with hypotenuse *AB* = 32, *M, N, P, Q,* and *R* are midpoints of segments *AB, AC, CB, BM, and AM*, respectively. Find the perimeter of *NPQR*.

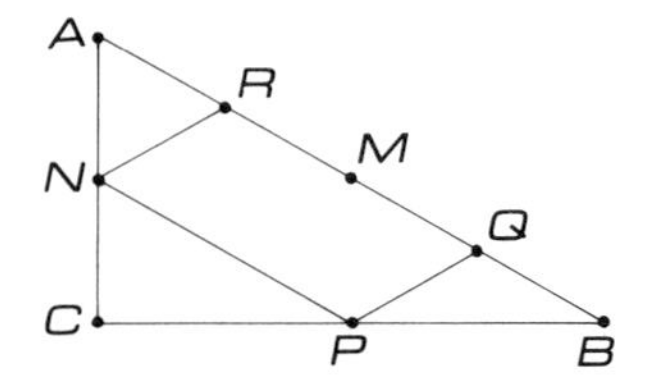

***Fig. 7.4.* Numerical context for logical reasoning**

In this single exercise, students would need to apply logical reasoning in translating the information into a diagram, to derive useful information from appropriate theorems, and to organize that information to lead to the correct calculations. The connection between the problem solution and the sequence of steps in a proof should be emphasized.

College-intending students also should gain an appreciation of Euclidean geometry as one of many axiomatic systems. This goal may be achieved by directing students to investigate properties of other geometries to see how the basic axioms and definitions lead to quite different—and often contradictory—results. For example, great circles, which play the role of lines in spherical geometry, always meet. Thus, in spherical geometry, instead of having *exactly one* line parallel to a given line through a point not on the line, there are *no* such lines. Figure 7.5 shows another interesting difference between Euclidean geometry and the geometry of a sphere. Students also could examine some of the history associated with attempts to prove Euclid's famous fifth postulate from both a mathematical and a cultural perspective.

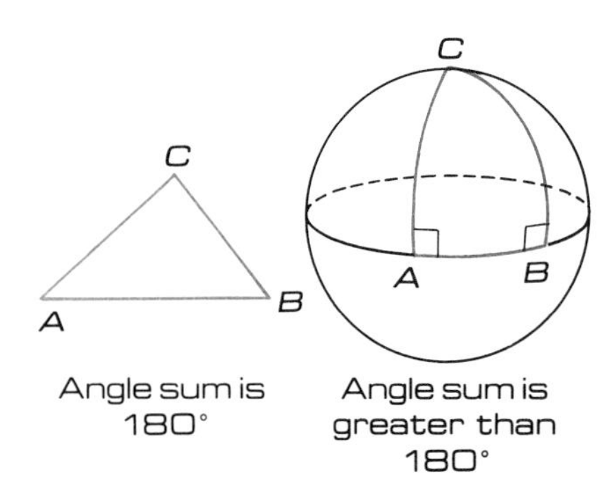

***Fig. 7.5.* Geometry of a sphere**

In summary, synthetic geometry at the high school level should focus on more than deductive reasoning and proof. Equally important is the continued development of students' skills in visualization, pictorial representation, and the application of geometric ideas to describe and answer questions about natural, physical, and social phenomena.

STANDARD 8: GEOMETRY FROM AN ALGEBRAIC PERSPECTIVE

In grades 9–12, the mathematics curriculum should include the study of the geometry of two and three dimensions from an algebraic point of view so that all students can—

- ***translate between synthetic and coordinate representations;***
- ***deduce properties of figures using transformations and using coordinates;***
- ***identify congruent and similar figures using transformations;***
- ***analyze properties of Euclidean transformations and relate translations to vectors;***

and so that, in addition, college-intending students can—

- ***deduce properties of figures using vectors;***
- ***apply transformations, coordinates, and vectors in problem solving.***

Focus

One of the most important connections in all of mathematics is that between geometry and algebra. Historically, mathematics took a great stride forward in the seventeenth century when the geometric ideas of the ancients were expressed in the language of coordinate geometry, thus providing new tools for the solution of a wide range of problems.

More recently, the study of geometry through the use of transformations—the geometric counterpart of functions—has changed the subject from static to dynamic, providing in the process great additional power that can be used, for example, to describe and produce moving figures on a video screen. Viewed as an algebraic system, transformations also provide college-intending students with valuable experiences with properties of function composition and group structure.

The interplay between geometry and algebra strengthens students' ability to formulate and analyze problems from situations both within and outside mathematics. Although students will at times work separately in synthetic, coordinate, and transformation geometry, they should have as many opportunities as possible to compare, contrast, and translate among these systems. A fundamental idea students should come to understand is that specific problems are often more easily solved in one or another of these systems.

Discussion

Objects and relations in geometry correspond directly to objects and relations in algebra. For example, a point in geometry corresponds to an ordered pair (x, y) of numbers in algebra, a line to a set of ordered pairs satisfying an equation of the form $ax + by = c$, and the intersection of two lines to the set of ordered pairs that satisfy the corresponding equations. It is correspondences like these that allow translation between the two "languages" and permit concepts in one to clarify and

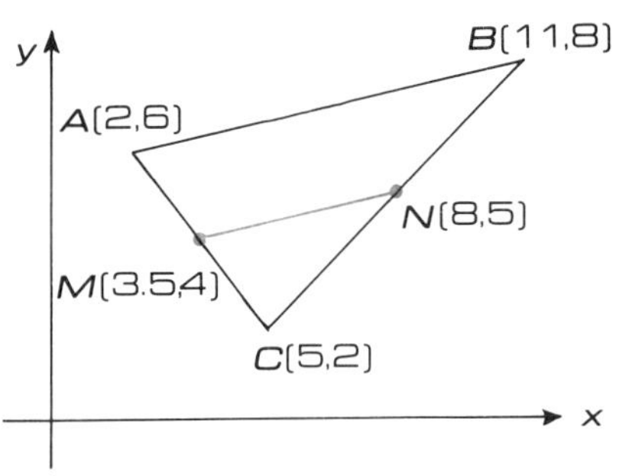

Students would use the midpoint formula to compute the coordinates of midpoints *M* and *N*. Using the distance formula, they establish that $MN = (1/2)AB$. Finally comparison of slopes of $\overline{MN}$ and $\overline{AB}$ demonstrate the parallelism.

***Fig. 8.1.* A coordinate argument**

reinforce concepts in the other. In fact, deducing properties of geometric figures using their coordinate representations is often easier for students than synthetic proofs are. For example, all students can calculate the coordinates of the midpoints of two sides of a triangle with given numerical coordinates and use the results to deduce that the segment joining them is parallel to the third side of the triangle and equals half its length (fig. 8.1). College-intending students should be able to extend this technique to prove the result in general.

Although students will continue to work with two dimensions, every opportunity should be taken to explore the third dimension as well. Algebraic formulations in three-dimensional coordinate geometry should focus on figures that are simple to represent, such as points, planes perpendicular to an axis, and spheres. The coordinate representation of general planes and lines is more difficult and would best be treated as an enrichment project.

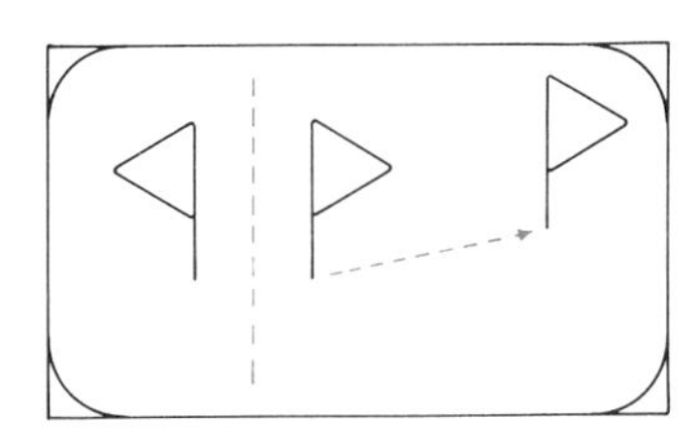

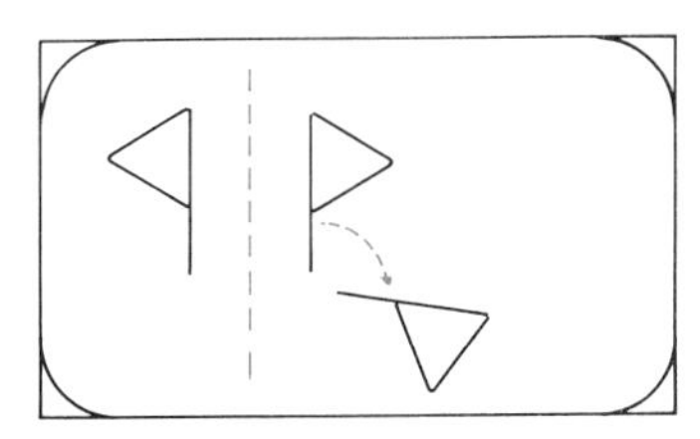

***Fig. 8.2.* Effects on the plane of the composites of two transformations**

Transformations serve as a powerful problem-solving tool and permit students to develop a broad concept of congruence and similarity that applies to all figures. The derivation of congruence properties through isometries (distance-preserving transformations) and of similarity properties through composites of dilations (ratio-preserving transformations) and isometries provides a connection with, and a reinforcement of, synthetic methods. Transformation geometry in three dimensions is straightforward: most three-dimensional transformations are simply direct extensions of their two-dimensional counterparts.

Transformations often are used to represent physical motions, such as slides, flips, turns, and stretches. Students should use computer software based on this dynamic view of transformations to explore properties of translations, line reflections, rotations, and dilations, as well as compositions of these transformations (fig. 8.2). These graphics experiences not only help students develop an understanding of the effects of various transformations but also contribute to the development of their skills in visualizing congruent and similar figures.

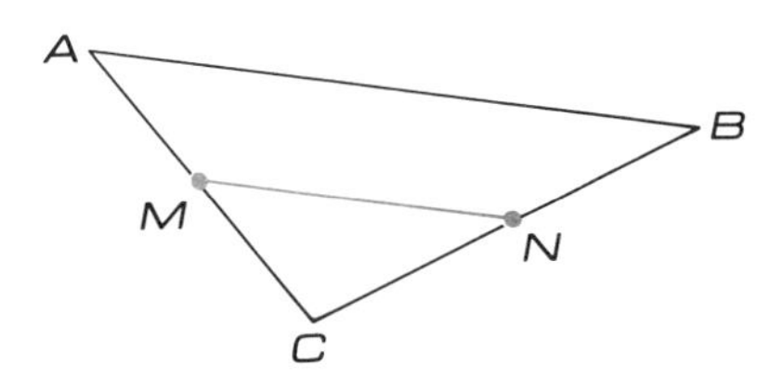

Under a dilation with center at *C* and scale factor 2, $\overline{MN} \rightarrow \overline{AB}$ so $AB = 2MN$ and $\overline{AB} \parallel \overline{MN}$

***Fig. 8.3.* A transformation proof**

The midpoint theorem discussed previously in terms of a coordinate approach also can be proved using a dilation with scale factor 2 (or 1/2) and with center at the triangle vertex that is common to the two sides whose midpoints are connected (see fig. 8.3).

Although this theorem can also be proved by synthetic methods, students should appreciate the economy offered by coordinate, transformation, and even vector techniques (see fig. 8.4).

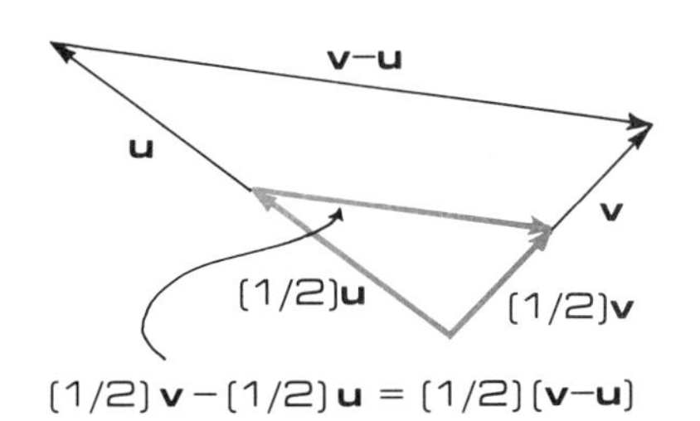

$(1/2)\,\mathbf{v} - (1/2)\,\mathbf{u} = (1/2)\,(\mathbf{v}-\mathbf{u})$

***Fig. 8.4.* A vector proof**

It is important that all students come to understand how vectors can be used to represent such physical phenomena as velocity and force. They should investigate vector addition and scalar multiplication, both geometrically and algebraically, and be introduced to applications of those ideas such as in quantifying the effect of wind on the course of an airplane. College-intending students should develop facility with the use of vectors to solve problems as well as to prove geometric theorems (as exemplified in fig. 8.4).

Taken together, the two geometry standards advocate a more eclectic approach to the subject, one based on informal explorations and short local axiomatic sequences. The study of geometry should provide students with the ability to recognize and apply effectively the geometric concepts and methods (synthetic, coordinate, transformation, or vector) most appropriate to a given problem situation.

STANDARD 9: TRIGONOMETRY

In grades 9–12, the mathematics curriculum should include the study of trigonometry so that all students can—

- ***apply trigonometry to problem situations involving triangles;***
- ***explore periodic real-world phenomena using the sine and cosine functions;***

and so that, in addition, college-intending students can—

- ***understand the connection between trigonometric and circular functions;***
- ***use circular functions to model periodic real-world phenomena;***
- ***apply general graphing techniques to trigonometric functions;***
- ***solve trigonometric equations and verify trigonometric identities;***
- ***understand the connections between trigonometric functions and polar coordinates, complex numbers, and series.***

Focus

Trigonometry has its origins in the study of triangle measurement. Many real-world problems, including those from the fields of navigation and surveying, require the solution of triangles. In addition, important mathematical topics, such as matrix representations of rotations, direction angles of vectors, polar coordinates, and trigonometric representations of complex numbers, require trigonometric ratios, further underscoring the connections between geometry and algebra.

Natural generalizations of the ratios of right-angle trigonometry give rise to both trigonometric and circular functions. These functions, especially the sine and cosine, are mathematical models for many periodic real-world phenomena, such as uniform circular motion, temperature changes, biorhythms, sound waves, and tide variations. Although all students should explore data from such phenomena, college-intending students should identify and analyze the corresponding trigonometric models. These students should also study identities involving trigonometric expressions and inverses of trigonometric functions, together with their applications to the solution of equations and inequalities.

Scientific calculators can and should significantly facilitate the teaching of trigonometry, providing more class time and computational power to develop conceptual understanding and address realistic applications. Graphing utilities provide dynamic tools that permit students to model many realistic problem situations using trigonometric equations or inequalities. Consistent with the other standards, graphing utilities also should play an important role in students' development of an understanding of the properties of trigonometric functions and their inverses. In addition, college-intending students should solve trigonometric equations and inequalities by computer-based methods, such as those described in the standard on algebra.

Discussion

All students should apply trigonometric methods to practical situations involving triangles. As an example, consider a right-triangle surveying problem with which cartographers are frequently confronted.

Determine the angle of depression between two markers on a contour map with different elevations.

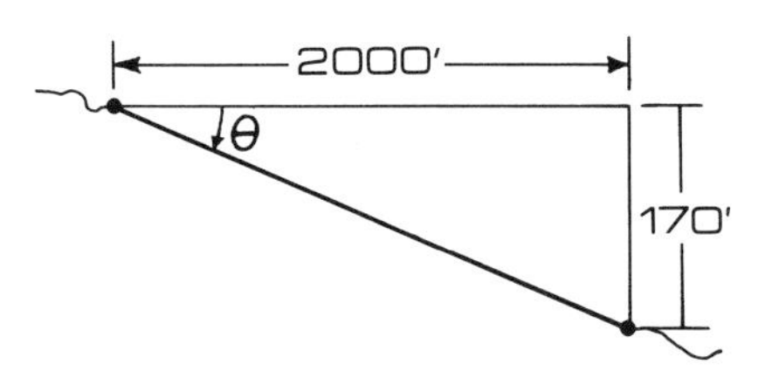

Fig. 9.1. A practical application of trigonometry

Students would first develop a geometric model (fig. 9.1) based on information read from the map. They would then identify a trigonometric ratio appropriate to the situation, write the corresponding equation, use a calculator to readily obtain a numerical answer, and then interpret this value to the appropriate degree of accuracy in terms of the given units of measure. College-intending students also should derive and apply the laws of sines and cosines to problem situations involving general triangles.

All students should use the sine and cosine functions to model periodic real-world phenomena. One setting with which the majority of students are familiar is that of a Ferris wheel.

Suppose a Ferris wheel with a radius of 25 feet makes a complete revolution in 12 seconds. Develop a mathematical model that describes the relationship between the height *h* of a rider above the bottom of the Ferris wheel (4 feet above the ground) and time *t*.

This problem can be addressed within the core curriculum by students at several possible levels of formalism.

Level 1: At this level, students would first develop a table of t- and h-values. Assuming that the rider is at the bottom of the Ferris wheel when $t = 0$, students can easily determine values of h for $t = 0, 3, 6, 9, 12$. For t-values between these numbers, values of h could be estimated from a scale drawing of the Ferris wheel as in figure 9.2(a). By plotting the collected data (fig. 9.2(b)) and noting the periodicity of the function, students may conjecture that the graph has a sinusoidal shape and thereby predict its shape for larger values of t.

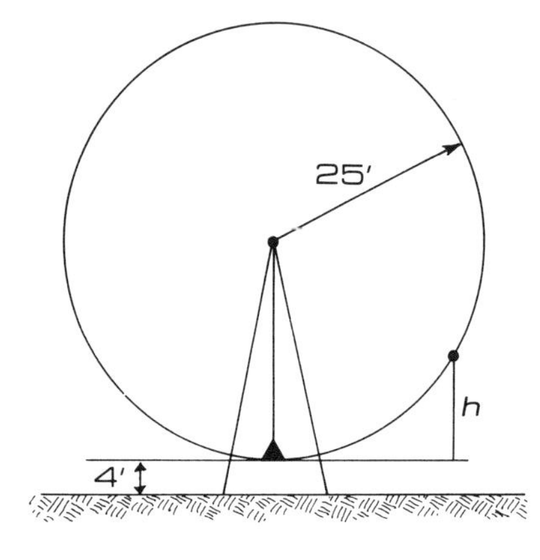

Time t (sec)	Height h (ft)	Height above ground (ft)
0	0	4
1.5	7	11
3	25	29
4.5	43	47
6	50	54
7.5	43	47
9	25	29
10.5	7	11
12	0	4

(a)

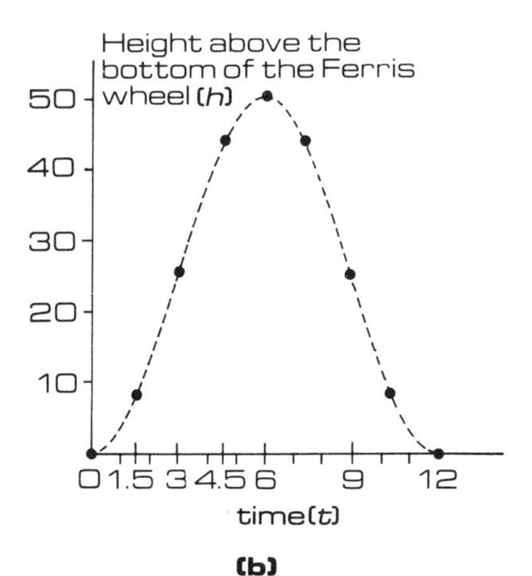

(b)

Fig. 9.2. Modeling the position of a rider on a Ferris wheel

Level 2: Students at this level would be given an equation for $h(t)$ (such as $h(t) = -25 \cos (\pi/6)t + 25$) and asked to graph it and then to analyze their graphs. The interpretation of their graphs should focus on the contextual meaning of the local maximum and minimum points, finding h-values for given t-values and t-values for given h-values, and finding the number of revolutions for some (large) t-value and the time t required for a given number of revolutions. Finally, students would explore the changes in the graph for a Ferris wheel that has a different radius or rate of revolution.

Level 3: Recognizing that the graph obtained through experiences such as those in level 1 is that of a function of the form $h(t) = a \cos(bt) + c$, students at this level would proceed to determine a, b, and c by comparing the graph of $f(t) = \cos t$ to their graph. This analysis would suggest the need to reflect the graph of f across the t-axis and then to adjust the amplitude, period, and shift in the vertical direction.

Level 4: At this level, students would use right-triangle trigonometry and simple proportions (see fig. 9.3) to derive the parametric representation of a point $P = (x(t), y(t))$ on the rotating Ferris wheel as a function of time, thereby *establishing* that the height is a sinusoidal function of t. They could then use a parametric graphing utility to simulate the motion of a point moving on the Ferris wheel.

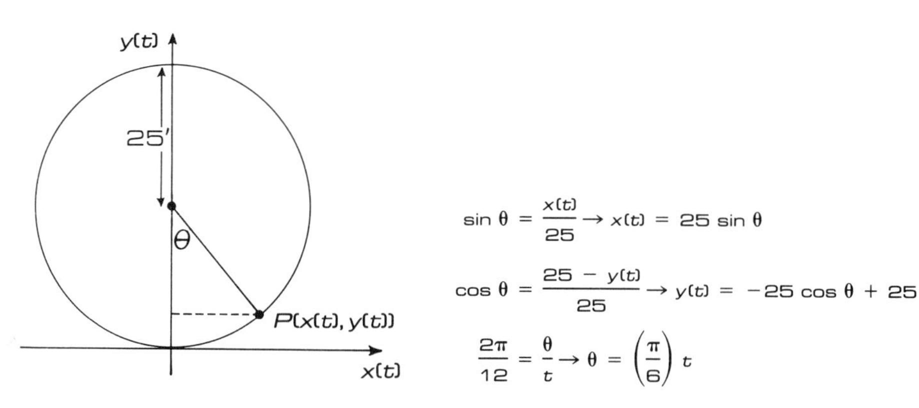

Fig. 9.3. Parametric representation of P

We again emphasize that the entry and exit levels with respect to the treatment of this, or any, particular topic is largely determined by the background of the students and their performance in the activity itself. For some students these two levels may be the same. Seldom would any student progress through all levels within a single unit of study.

Concepts related to trigonometric functions such as amplitude, period, and phase shift should be introduced to college-intending students through real-world applications. These students will have had experience with graphs of functions of the form $y = af(bx + c) + d$, including the investigation of the effects of changing the parameters a, b, c, and d on the graph of $y = f(x)$. Thus, after appropriate computer-graphing experiences, they should be able to sketch quickly, without the aid of a computer, the graph of a function like $y = 3 \sin(x + 2)$ by applying two transformations to the graph of $y = \sin x$.

College-intending students also should have opportunities to verify basic trigonometric identities, such as $\sec^2 A = 1 + \tan^2 A$, since this activity improves their understanding of trigonometric properties and provides a new setting for deductive proof. Only minimal amounts of class time should be devoted to verifying identities, however, and artificially complicated identities, such as $\csc^6 x - \cot^6 x = 1 + 3 \csc^2 x \cot^2 x$, should be avoided altogether.

College-intending students should also develop an understanding of the connections between trigonometric functions and the topics of polar coordinates, complex numbers, and series. Using a calculator or a computer, for example, students can investigate the power-series expansion of the sine function numerically and graphically. Figure 9.4 illustrates how the first five terms of the series expansion for the sine function very closely approximate the values of the sine function for $|x| \leq 4$.

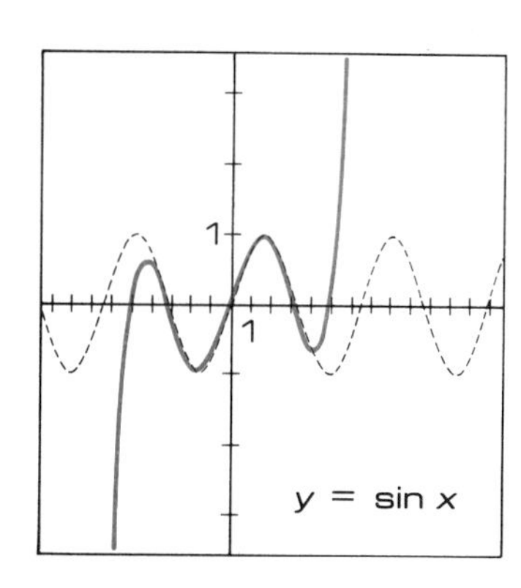

$y = x - (1/3!)x^3 + (1/5!)x^5 - (1/7!)x^7 + (1/9!)x^9$

Fig. 9.4. Series expansion of the sine function

Students could use a graphing utility to explore such issues as the number of terms of the expansion necessary for the series to closely approximate the sine function for $|x| \leq 10$. Such an approach can lead to valuable discussions of limits and errors in approximations.

Trigonometry not only remains an important and powerful tool for science and engineering but also continues to provide an aesthetic attraction for many students through its regularities and symmetries. Calculator and computer technology makes both aspects of the subject readily accessible to a wider range of students and at an earlier age level. This in turn provides opportunities for greater integration of trigonometry with geometry and algebra.

♦ ♦ ♦ ♦ ♦ ♦ ♦ ♦

STANDARD 10: STATISTICS

In grades 9–12, the mathematics curriculum should include the continued study of data analysis and statistics so that all students can—

- ***construct and draw inferences from charts, tables, and graphs that summarize data from real-world situations;***
- ***use curve fitting to predict from data;***
- ***understand and apply measures of central tendency, variability, and correlation;***
- ***understand sampling and recognize its role in statistical claims;***
- ***design a statistical experiment to study a problem, conduct the experiment, and interpret and communicate the outcomes;***
- ***analyze the effects of data transformations on measures of central tendency and variability;***

and so that, in addition, college-intending students can—

- ***transform data to aid in data interpretation and prediction;***
- ***test hypotheses using appropriate statistics.***

Focus

Collecting, representing, and processing data are activities of major importance to contemporary society. In the natural and social sciences, data are also summarized, analyzed, and transformed. These activities involve simulations and/or sampling, fitting curves, testing hypotheses, and drawing inferences. To enhance their social awareness and career opportunities, students should learn to apply these techniques in solving problems and in evaluating the myriad statistical claims they encounter in their daily lives.

The study of statistics in grades 9–12 should consolidate, deepen, and build on student understandings of methods of exploratory data analysis as developed in the elementary and middle grades. Students should be encouraged to apply statistical tools to other academic subjects through the exploration of such data as student-opinion polls for social studies, word or letter counts for English, and plant-growth records for biology. Out-of-school activities such as athletics provide further opportunities for data analysis, the results of which can be seen to be immediately useful.

It is essential that students come to understand the difference between the right-or-wrong quality characteristic of most mathematical thinking and the qualified nature of outcomes in statistical analysis. It is equally important, however, that students do not extrapolate beyond this fact to reject statistical thinking because it allows counterexamples. Instead, they should recognize that statistics plays an important intermediate role between the exactness of other mathematical studies and the equivocal nature of a world dependent largely on individual opinion.

Computing technology allows students to represent data in graphs quickly (with curve fitting done for them) and to calculate statistical measures with remarkable precision using single computer keystrokes. What is missing—and what their study of statistics should provide—is an understanding of which measures are appropriate for a given problem and what such measures as mean, variance, and correlation can tell them about a problem. Furthermore, it is essential that students learn to interpret results intelligently.

Discussion

This standard should not be viewed as advocating, or even prescribing, a statistics course; rather, it describes topics that should be integrated with other mathematics topics and disciplines. For example, curve fitting is a statistical topic that integrates easily into the study of linear and higher-order equations. Students could investigate the possible relationship between car age and mileage by collecting data from the school parking lot and constructing a scatter plot (fig. 10.1).

Model Year	Mileage
'68	80 000
'71	168 000
'72	112 000
'74	44 000
'76	141 000
'77	96 000
'78	138 000
'79	120 000
'82	72 000
'83	69 000
'83	68 000
'84	93 000
'84	37 000
'85	47 000
'85	45 000
'86	29 000
'86	28 000
'86	27 000
'87	18 000
'87	17 000
'87	16 000

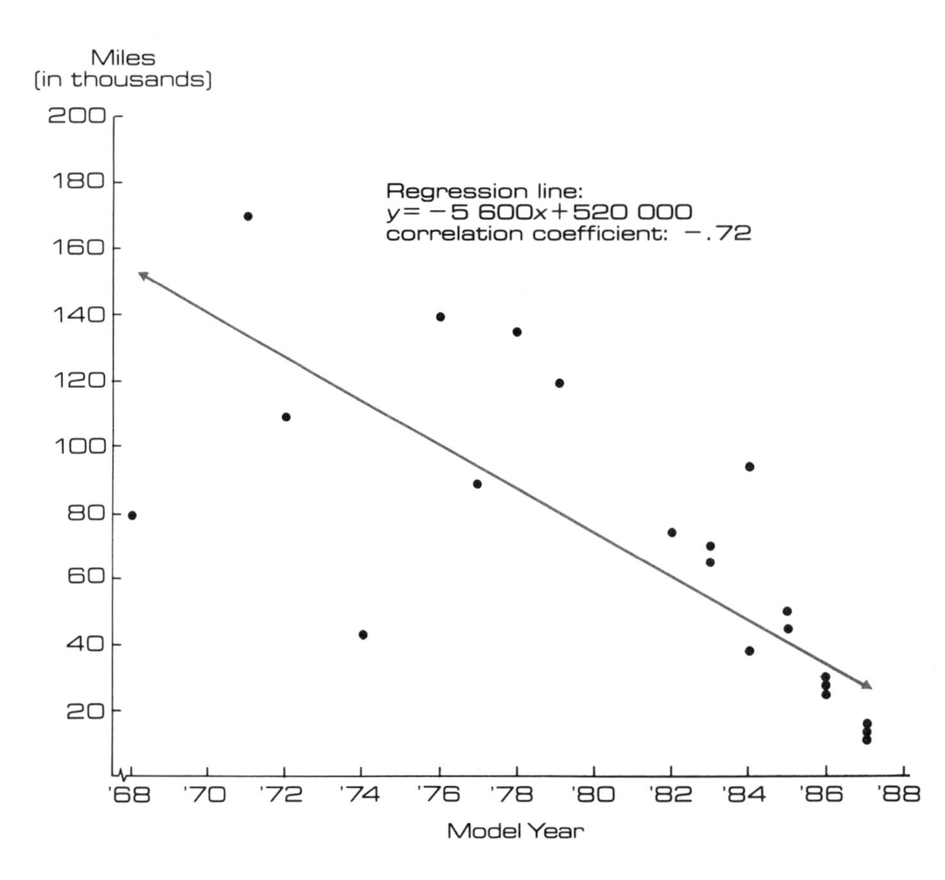

Fig. 10.1. Car mileage by model year

Since the points seem to lie in a reasonably narrow band, students can identify a "best" line that fits their data. Techniques for constructing such a line can range from the most basic, such as placing a piece of uncooked spaghetti or a string on the graph so that approximately the same number of data points fall above as below, to approaches involving medians of grouped data or the technique of least squares. Many calculators provide the capability to generate an equation for the regression line (one of these "best" lines) as well as the associated correlation coefficient. Students then can use either their graph or their equation to predict, for example, the expected mileage of a 1980 car. They should be encouraged to write a summary paragraph about the information displayed in the table or graph and include inferences they believe are supported by their analysis of the data.

♦ ♦ ♦ ♦ ♦ ♦ ♦ ♦

Students should also come to realize that curve fitting is not appropriate for all data sets. In paperback books, for example, there is so little relationship between the number of pages and the price that further analysis would not yield useful information.

Communication plays a central role in statistical problems. Quantitative results require careful exposition and interpretation if they are to have meaning. In particular, it is often true that different modes of data representation convey quite different messages. A regression line (calculated directly from data without reference to a scatter plot) might be strongly influenced by a few aberrant points, for example, whereas the scatter plot for the same data might suggest that these outliers represent anomalies that may be due to mistakes in data handling. A further investigation of these specific points might lead to their rejection, a new curve fit, and an improved correlation.

Students must acquire intuitive notions of randomness, representativeness, and bias in sampling to enhance their ability to evaluate statistical claims. These understandings would give students the appropriate tools for rejecting such television advertising claims as one that portrays a series of people choosing the same commercial toothpaste. (Here the implication of representativeness clearly is not fulfilled.) If they are to grasp the concepts of sampling, the central limit theorem, and confidence intervals, students should have experience constructing and analyzing sampling distributions through simulations. These experiences provide students with the tools and the perspective they need to interpret such claims in the media as, "The polls indicate that 55% of the voters, with an error of 3%, prefer candidate X (with 90% confidence)." Statistics and probability threads are interwoven here. College-intending students also should apply this understanding of sampling in designing their own experiments to test hypotheses.

Students should be aware that bias can arise in the interpretation of results as well as in sampling: the interpreter's predisposition or expectation may strongly affect the message derived from the statistical results. This often occurs in the presentation and interpretation of data gathered for political purposes.

College-intending students should become familiar with such distributions as the normal, Student's t, Poisson, and chi square. Students should be able to determine when it is appropriate to use these distributions in statistical analysis (e.g., to obtain confidence intervals or to test hypotheses). Instructional activities should focus on the logic behind the process in addition to the "test" itself.

In recent years, nonparametric methods or distribution-free methods like the chi-square test in the cola example in the standard on probability (Standard 11) have increasingly been used as alternatives to statistical tests that assume a particular (often normal) distribution. Nonparametric techniques (which also include such measures as the sign test, the Mann-Whitney U test, and Spearman's rank correlation test) are extremely versatile, easy to use, often derive their power directly from combinatorics and the binomial distribution (of the statistics, not the sample), and are particularly well suited to small samples. As these methods continue to gain in popularity, it is expected that they will become an integral part of the evolving statistics curriculum.

All students should be encouraged to discover generalizations that relate the effect of modifying a set of data by addition or scalar multiplication on the mean,median, mode, and variance. For example, class test scores could be transformed by increasing each score by 10 points (or by multiplying each score by 1.1). Technology provides an easy means by which students can compute the statistics for the transformed data, which, on analysis, lead to generalizations that the mean, median, and mode are increased by 10 (multiplied by 1.1) and the variance is unchanged (multiplied by $(1.1)^2$). College-intending students should be able to derive these results algebraically.

Statistical data, summaries, and inferences appear more frequently in the work and everyday lives of people than any other form of mathematical analysis. It is therefore essential that all high school graduates acquire, at the appropriate level, the capabilities identified in this standard. This expectation will require that statistics be given a more prominent position in the high school curriculum.

STANDARD 11: PROBABILITY

In grades 9–12, the mathematics curriculum should include the continued study of probability so that all students can—

- ***use experimental or theoretical probability, as appropriate, to represent and solve problems involving uncertainty;***
- ***use simulations to estimate probabilities;***
- ***understand the concept of a random variable;***
- ***create and interpret discrete probability distributions;***
- ***describe, in general terms, the normal curve and use its properties to answer questions about sets of data that are assumed to be normally distributed;***

and so that, in addition, college-intending students can—

- ***apply the concept of a random variable to generate and interpret probability distributions including binomial, uniform, normal, and chi square.***

Focus

Probability provides concepts and methods for dealing with uncertainty and for interpreting predictions based on uncertainty. Probabilistic measures are used to make marketing, research, business, entertainment, and defense decisions, and the language of probability is used to communicate these results to others. In grades 9–12, students should extend their K–8 experiences with simulations and experimental probability to continue to improve their intuition. These experiences provide students with a basis of understanding from which to make informed observations about the likelihood of events, to interpret and judge the validity of statistical claims in view of the underlying probabilistic assumptions, and to build more formal concepts of theoretical probability.

Discussion

Students in grades 9–12 should understand the differences between experimental and theoretical probability. Concepts of probability, such as independent and dependent events, and their relationship to compound events and conditional probability should be taught intuitively. Formal definitions and properties should be developed only after a firm conceptual base is established so that students do not apply formulas indiscriminately when solving probability problems. At this level, the focus of instructional time should be shifted from the selection of the correct counting technique to analysis of the problem situation and design of an appropriate simulation procedure. Some students also should be encouraged to approach problems from the perspective of a theoretical model.

It is also important for students to understand the differences between, and the advantages associated with, theoretical and simulation techniques. Even more important, students should value both approaches. For example, students might obtain a result through the application of a theoretical model and validate that result through simulation. What should *not* be taught is that only the theoretical approach yields the "right" solution.

Although probability provides useful models for the solution of problems in fields such as medicine, physics, and economics, many of the problems in daily living also can be better understood from this perspective.

Suppose Anne tells you that under her old method of shooting free throws in basketball, her average was 60%. Using a new method of shooting, she scored 9 out of her first 10 throws. Should she conclude that the new method really is better than the old method?

This problem can be addressed within the core curriculum at increasing levels of abstraction.

All students should first identify the real (statistical) question: What are the chances of shooting at least nine out of ten if you normally shoot 60%?

Level 1: Students model the problem by associating baskets with the integers 4–9 inclusive and misses with the numbers 0–3 inclusive. They then roll a fair icosahedral die (twenty faces with the digits 0–9 appearing twice) ten times. If nine or more "baskets" occur, count the trial as a success. For the first trial (see fig. 11.1), the digits 4–9 occur only seven times, so this trial is not a success. Students repeat the experiment nine more times and determine the percentage of successes.

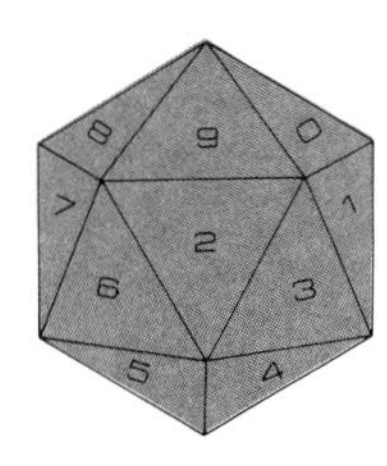

Icosahedral Die

Trials

1	2	3	4	5	6	7	8	9	10
1	6	8	7	7	0	4	4	1	9
7	5	9	3	9	9	5	5	6	7
7	6	4	3	1	7	5	4	0	7
5	5	7	6	2	4	1	4	0	3
5	9	2	6	9	3	8	7	9	3
4	4	7	2	2	7	6	7	9	2
5	4	1	4	6	1	3	2	5	8
1	0	0	1	7	5	9	6	5	7
4	1	3	8	9	4	1	8	5	8
1	9	5	6	6	8	3	5	0	3

***Fig. 11.1.* Free-throw shooting simulation**

For this set of ten trials, there is only one success—trial 8. So the chances of making at least nine out of ten shots is 1/10, or 10%. To obtain a better estimate of this probability, the results from the class would be pooled.

Level 2: Students use a random-number table (fig. 11.2) for the same kind of simulation. The digits 0, 1, 2, 3 correspond to misses; the digits 4, 5, 6, 7, 8, 9 to baskets. Each trial consists of selecting ten digits; a success (nine or ten baskets in ten free throws) occurs if the digits 0, 1, 2, 3 occur less than twice in each set of ten. For the trial shown, there are four misses and six baskets; hence the trial is not a success.

32236	12683	41949	91807	57883	65394	35595	39198	75268
40336	(50658	32089)	78007	58644	73823	62854	31151	64726
88795	93736	22189	47004	48304	77410	78871	98387	44647
12807	65194	58586	78232	57097	01430	00304	32036	23671
62932	99837	20160	27792	37090	62165	11172	66827	39830

***Fig. 11.2.* Selection of ten uniformly distributed random digits**

♦ ♦ ♦ ♦ ♦ ♦ ♦ ♦

Level 3: Students develop and use a computer program with a random-number generator to run the experiment 1000 times and compute the ratio of successful trials.

```
REM *** This program runs 1000 10-shot experiments shooting free
throws

Successes = 0
FOR trial = 1 to 1000
    Counter = 0
    FOR shotnumber = 1 to 10
            X = INT(10*RND)
            IF X > = 4 THEN Counter = Counter + 1
    NEXT shotnumber
    IF Counter > = 9 THEN Successes = Successes + 1
NEXT trial
PRINT "The probability of shooting at least 9 baskets is";
Successes/1000

>RUN

>The probability of shooting at least nine baskets is .048.
```

Level 4: Students relate the problem to the binomial distribution and the specific question to the binomial probabilities (10, 9; 0.6) and (10, 10; 0.6).

The probability of success: probability of making exactly nine shots + probability of making all ten shots

$$P\ (\text{success}) = ({}_{10}C_9)(.6)^9(.4)^1 + ({}_{10}C_{10})(.6)^{10}(.4)^0$$
$$= 10(.6)^9(.4)^1 + (.6)^{10}$$
$$\doteq .046$$

On completing work at any one of the levels, students would be asked to interpret their obtained probabilities and formulate a response to Anne about her new method of shooting free throws.

The reader should note that although the problem setting considered here remains the same for all students, the instructional treatment varies not only in the content and its complexity of language and notation but also in the type of resources (polyhedral dice, random-number table, computer-based random-number generation) used. The impact of computing technology on the teaching of probability can be clearly seen by comparing the results obtained by informal simulation methods (level 3) with those obtained by formal analytic methods (level 4). Even students who successfully completed work at levels 1 or 2 could discuss and then use a computer program based on their model.

Once students have acquired an intuitive understanding of a random variable, they can extend the concept of the probability of an event to the development of a probability distribution. For example, students could associate the outcome of each roll of two dice with the product of the numbers on each face as illustrated in figure 11.3.

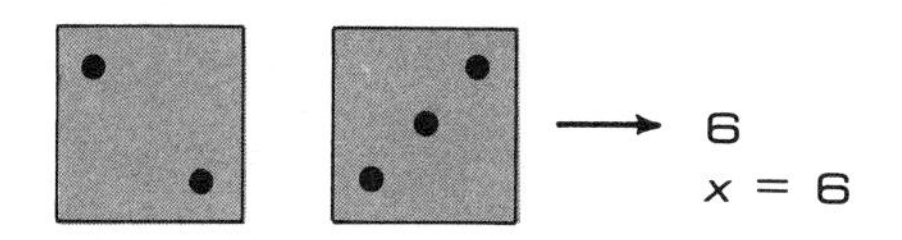

Fig. 11.3. Introducing the concept of a random variable

Probabilities now can be assigned to each outcome, experimentally by simulation or theoretically by constructing a table of outcomes. This can be expressed by college-intending students in formal notation, such as

$P(x = 6) = 1/9$, where x denotes the random variable. All students can construct a graph of this discrete probability distribution to answer such questions as, What is the probability of obtaining a product greater than 10?

In addition to generating discrete probability distributions, such as the binomial $(n, x; p)$ for various values of n and p, college-intending students could use a mathematical statement for a random variable (e.g., $X =$ INT(−5*log(RND)), where INT is the greatest-integer function and RND represents a random number, $0 < \text{RND} < 1$, and construct the associated frequency distribution (in this example, Poisson). This kind of distribution occurs in such situations as queues and traffic flow.

Furthermore, college-intending students could also relate the coefficients of the terms in the expansion of a power of a binomial expression—for example, $(p + q)^n$—to the frequency of events in a binomial distribution. Finite random walks on a number line or in the first quadrant are appropriate applications of these concepts and can easily be simulated on a computer.

Because the distributions of data from many real-world phenomena can be closely approximated by the normal curve, all students should become familiar with the geometric properties of its graph and should be able to use either probability tables associated with the curve or computer software to solve problems. Here is one example:

Assuming that gas consumption for a car model is normally distributed with a mean of 25.7 mpg and a standard deviation of 2.9 mpg, how likely would it be that a particular car of this model gets at least 30 mpg?

To solve this problem, it is first necessary to standardize the random variable (gas consumption) using transformations on the mean and standard deviation as discussed in the standard on statistics.

Here $P(\text{mpg} \geqslant 30) = P(z \geqslant 1.48)$. In figure 11.4, the area of the shaded region corresponds to the probability of this event. From a standardized normal-curve table, this probability is found to be approximately .07. College-intending students should also investigate similarities between the binomial and normal distributions.

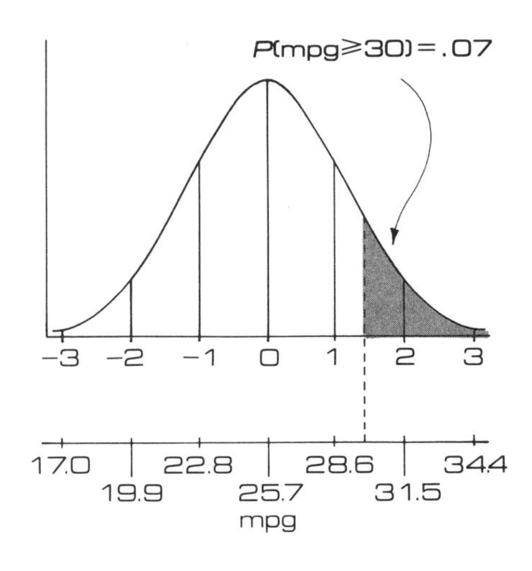

***Fig. 11.4.* Gas mileage distribution**

Probability and statistics should be developed in a manner that highlights their interrelatedness. If students are to test hypotheses in statistics, they need a good understanding of probability distributions. As an illustration, consider the following problem:

In a taste-test experiment involving three types of cola, 30 people make a selection: 15 choose Brand X, 8 Brand Y, and 7 Brand Z. The manufacturer of Brand X claims that these data show its superiority over the others. Is his claim reasonable? How likely is it that the outcome could have occurred by chance if there were no preference?

In solving this problem, students can develop some new mathematics and in the process gain understanding of the chi-square statistic. Here, in capsule form, is one process they might follow:

1. Students agree that a 10-10-10 distribution of choices would result if there were no preference.

2. Deviations from this result can be handled so that they won't cancel each other out, by computing $(10 - 15)^2$, $(10 - 8)^2$, and $(10 - 7)^2$, as in a variance calculation.

◆ ◆ ◆ ◆ ◆ ◆ ◆ ◆

3. In order to scale these values so that they may be compared with similar distributions, they each are divided by the expected value and then summed. The resulting value is 3.8, which is a particular value of a random variable called a *chi-square statistic* and is denoted χ^2.

4. The problem now becomes finding $P(\chi^2 \geqslant 3.8)$. Students may explore this probability through simulation of the cola experiment assuming no preference and tabulating the resulting frequency distribution of this random variable, χ^2, as in the following stem-and-leaf table for sixty trials (fig. 11.5). In this table, there are eight values for which $\chi^2 \geqslant 3.8$; thus, the associated $P(\chi^2 \geqslant 3.8) = 8/60$, or $.13^+$. Thus, the 15-8-7 distribution (or worse) would have occurred about 13% of the time even if there were no preference among the colas in a sample of thirty people.

5. Subsequent discussion of this result should convince students that a claim of preference for Brand X is not very strong.

In addition to providing information related to a specific (statistics) problem, these results also may be used to confirm the theoretical value to be found in a table for the chi-square statistic. This latter activity falls clearly within the realm of probability.

STEM	LEAF
0.	2222222222222
	66666666666668888
1.	44444444448888
2.	44666
3.	222888
4.	22
5.	6
6.	
7.	28

***Fig. 11.5.* Stem-and-leaf table for simulated cola experiments**

◆ ◆ ◆ ◆ ◆ ◆ ◆ ◆

STANDARD 12: DISCRETE MATHEMATICS

In grades 9–12, the mathematics curriculum should include topics from discrete mathematics so that all students can—

- ***represent problem situations using discrete structures such as finite graphs, matrices, sequences, and recurrence relations;***
- ***represent and analyze finite graphs using matrices;***
- ***develop and analyze algorithms;***
- ***solve enumeration and finite probability problems;***

and so that, in addition, college-intending students can—

- ***represent and solve problems using linear programming and difference equations;***
- ***investigate problem situations that arise in connection with computer validation and the application of algorithms.***

Focus

As we move toward the twenty-first century, information and its communication have become at least as important as the production of material goods. Whereas the physical or material world is most often modeled by continuous mathematics, that is, the calculus and prerequisite ideas from algebra, geometry, and trigonometry, the nonmaterial world of information processing requires the use of discrete (discontinuous) mathematics. Computer technology, too, wields an ever-increasing influence on how mathematics is created and used. Computers are essentially finite, discrete machines, and thus topics from discrete mathematics are essential to solving problems using computer methods. In light of these facts, it is crucial that all students have experiences with the concepts and methods of discrete mathematics.

Although discrete mathematics is a relatively new term, we will consider it simply to be the study of mathematical properties of sets and systems that have a countable number of elements.

Discussion

This standard neither advocates nor describes a separate course in discrete mathematics at the secondary school level; rather, it identifies those topics from discrete mathematics that should be integrated throughout the high school curriculum. The topics recommended were not selected exclusively because of their potential applications to computer science but because they represent useful mathematical ideas that have assumed increasing importance for all students. The depth and formalism of treatment should be consistent with the level of the courses in which a topic appears.

An emphasis on mathematics as a powerful representation tool pervades all the K–12 standards. Finite graphs (structures consisting of vertices and edges), together with their associated matrix representations, offer an important addition to the student's repertoire of representation schemes. For example, a complex network of one-way streets can be represented geometrically by a directed graph, which in turn can be interpreted algebraically as a matrix. An i-j entry in the matrix is 1 if and only if corresponding vertices are adjacent (i.e., connected by an edge); otherwise, the entry is 0. The general technique is illustrated by the simplified example in figure 12.1.

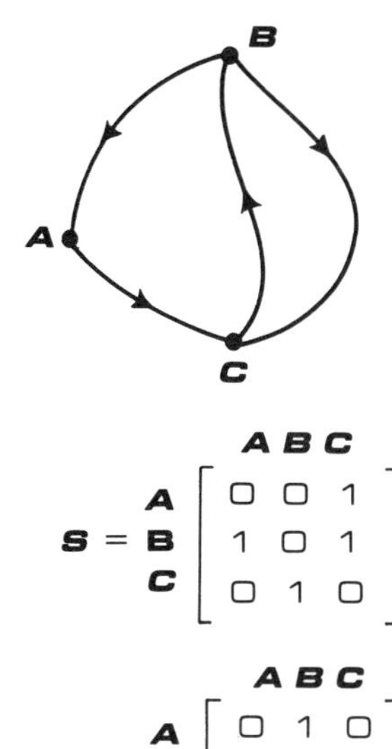

$$S = \begin{array}{c} \\ A \\ B \\ C \end{array} \begin{array}{c} A\ B\ C \\ \begin{bmatrix} 0 & 0 & 1 \\ 1 & 0 & 1 \\ 0 & 1 & 0 \end{bmatrix} \end{array}$$

$$S^2 = \begin{array}{c} \\ A \\ B \\ C \end{array} \begin{array}{c} A\ B\ C \\ \begin{bmatrix} 0 & 1 & 0 \\ 0 & 1 & 1 \\ 1 & 0 & 1 \end{bmatrix} \end{array}$$

Fig. 12.1. A graph and its adjacency matrix

By representing the graph as a matrix S and then multiplying S by itself, students can use S^2 to determine the number of two-stage routes connecting the various pairs of points. Students can generalize this procedure to graphs of any size, and computer software can be used to compute powers of the corresponding route matrices, which can then be analyzed to determine numbers of multiple-stage routes as well as other characteristics of networks.

Within the context of both the geometry and the algebra strands, students should have numerous opportunities to investigate this type of problem situation as well as those involving communication networks, circuit diagrams, tournament matchings, production schedules, and mathematical relations (e.g., the relation "is a factor of" defined on a finite set of integers), which can be effectively modeled and analyzed by graphs. Special graphs, called trees, are frequently used in the solution of probability problems and in representing the order of operations in algebraic expressions.

Sequences and series, currently topics in contemporary advanced algebra courses, should receive more attention, with a greater emphasis on their descriptions in terms of recurrence relations (formulas expressing each term as a function of one or more of the previous terms). Students can use recurrence relations to model real-world phenomena. The Fibonacci sequence is one model that has many applications:

1, 1, 2, 3, 5, 8, 13, 21, ...
Here, $a_0 = 1$, $a_1 = 1$, and $a_n = a_{n-1} + a_{n-2}$, for $n \geqslant 2$.

The numbers in the Fibonacci sequence occur surprisingly often in nature. They may be found in the arrangement of leaves around the branches of various trees, scales on pinecones, and the whorls on a pineapple. The analysis of these arrangements would provide an ideal setting for integrating the study of mathematics and botany.

Consumer applications essential for daily living should be included in the curriculum for all students. Recurrence relations can be used to provide a unified approach to such topics as compound interest, home mortgages, and annuities. For example, the compound-interest equation (see equation 1 on p. 133) can be generalized and expressed recursively as

$$A_0 = 100 \text{ and } A_n = A_{n-1}(1.06) \text{ for } n \geqslant 1.$$

The method of thinking recursively applies equally well to many geometric settings. For example, to determine the maximum number of regions into which a plane is separated by n lines, no two of which are parallel and no three of which intersect at a point, students should observe the following—

- For 0 lines, there is 1 region; that is, $r_0 = 1$.
- If there are r_{n-1} regions for $n - 1$ lines (fig. 12.2), when the nth line, l, is drawn, an additional region is formed as l intercepts each of the $n - 1$ lines. One more region is formed as l continues past the last line. Thus, $r_n = r_{n-1} + (n - 1) + 1 = r_{n-1} + n$, for $n \geqslant 1$.

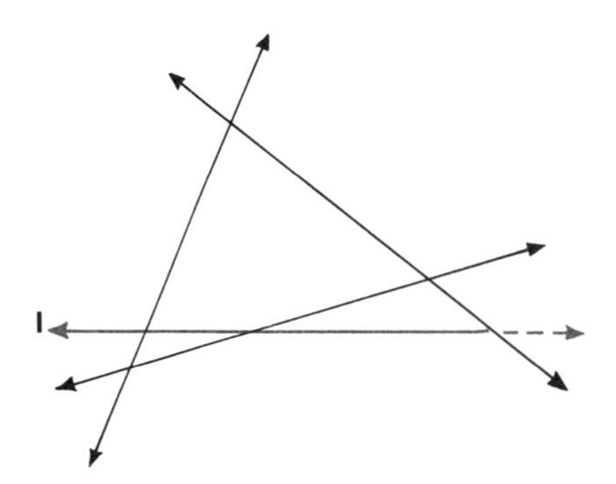

Fig. 12.2. Counting regions using recursive thinking

The recurrence relation can be used to solve the problem for any specified n. It also could be used to generate a sequence to which the method of finite differences could be applied to obtain a polynomial representation of the solution, in this case, $r = (1/2)n^2 + (1/2)n + 1$.

Instruction should stress the role of recurrence formulas as important tools for solving enumeration problems, since they can be translated easily to computer programs to obtain solutions. College-intending students should have the opportunity to use difference-equation techniques to express recurrence relations in closed form, that is, the nth term written as a function of n.

The development and analysis of algorithms lie at the heart of computer methods of solving problems. Thus, a consistent effort should be made throughout the 9–12 curriculum to provide opportunities for students to construct mathematics from an algorithmic point of view. Students must be encouraged to develop and analyze their own algorithms, not merely carry out those that have been prescribed to them. For example, approaching polynomial evaluation from an algorithmic point of view might involve the student in the following:

1. Verifying that $p(x) = a_0x^4 + a_1x^3 + a_2x^2 + a_3x + a_4$ can also be expressed in the form $p(x) = (((a_0x + a_1)x + a_2)x + a_3)x + a_4$

2. Designing an algorithm for evaluating $p(x)$ using the nesting pattern above (a possible notation for expressing the algorithm is shown below):

   Input x
   $p \leftarrow a_0$
   For $k = 1, 2, 3, 4$
   $\quad p \leftarrow p \cdot x + a_k$
   Output p

3. Comparing the number of operations used to evaluate p by the algorithm with the number needed when p is evaluated in standard form

4. Generalizing the algorithm so that it will evaluate a polynomial of degree n and then generalizing the answer in (3)

5. Engaging in a class discussion of the efficiency of the algorithm

Many of the topics in the secondary school mathematics curriculum can and should be investigated and developed by students from an algorithmic perspective. Possibilities include the greatest common factor of two integers; the solution of quadratic equations; approximating roots of polynomial equations; geometric constructions; the specification of a sequence of transformations mapping one figure onto another, similar one; the construction of Logo procedures to produce figures satisfying certain conditions; determining shortest/critical paths in finite graphs; random-number generation to simulate probability problems; sums of sequences; and solutions of systems of linear (and nonlinear) equations. The development and analysis of algorithms often add clarity and precision to the student's understanding of mathematical ideas and provide a context for nurturing careful logical reasoning.

In grades K–8, counting typically involves matching the elements of a set with a finite subset of the natural numbers. But real-world problems that can be simplified to the form "How many different subsets of size k can be selected from the members of a set having n distinct members?" re-

◆ ◆ ◆ ◆ ◆ ◆ ◆ ◆

quire an entirely different method of counting. To develop students' ability to solve problems with this structure, instruction should emphasize combinational reasoning as opposed to the application of analytic formulas for permutations and combinations. To illustrate this shift in perspective, consider a fundamental identity involving binomial coefficients: ${}_nC_r = {}_nC_{n-r}$. This identity usually is established by algebraic manipulation of the formula for combinations. In contrast, a student who reasons combinatorially may observe that ${}_nC_r$ represents the number of ways one can choose an r-element set from an n-element set. For each r-element set chosen, however, there corresponds a set of $n - r$ elements not chosen. Thus, the number of ways of choosing an r-element set is equal to the number of ways of choosing an $(n - r)$-element set.

Just as the topics of sequences, series, and recurrence formulas provide opportunities for reviewing algebra skills in a new context, so the suggested treatment of linear programming for college-intending students provides an opportunity to review topics from both geometry and algebra in a fresh context, rich in real-world applications.

From a discrete mathematics perspective, many topics now treated in contemporary courses and cited in other standards should receive increased attention: sets and relations; deductive proof, particularly proof by mathematical induction; the algebra of matrices; recursively defined functions; mathematical modeling; and algebraic structure.

STANDARD 13: CONCEPTUAL UNDERPINNINGS OF CALCULUS

In grades 9–12, the mathematics curriculum should include the informal exploration of calculus concepts from both a graphical and a numerical perspective so that all students can—

- ***determine maximum and minimum points of a graph and interpret the results in problem situations;***
- ***investigate limiting processes by examining infinite sequences and series and areas under curves;***

and so that, in addition, college-intending students can—

- ***understand the conceptual foundations of limit, the area under a curve, the rate of change, and the slope of a tangent line, and their applications in other disciplines;***
- ***analyze the graphs of polynomial, rational, radical, and transcendental functions.***

Focus

This standard does *not* advocate the *formal* study of calculus in high school for all students or even for college-intending students. Rather, it calls for opportunities for students to systematically, but informally, investigate the central ideas of calculus—limit, the area under a curve, the rate of change, and the slope of a tangent line—that contribute to a deepening of their understanding of function and its utility in representing and answering questions about real-world phenomena.

Most of the mathematics described in the other 9–12 standards involve *finite* processes, such as determining a sequence of transformations that maps a figure onto a congruent figure or approximating a zero of a polynominal function using an iterative technique. In contrast, the concept of limit and its connection with the other mathematical topics in this standard is based on *infinite* processes. Thus, explorations of the topics proposed here not only extend students' knowledge of function characteristics but also introduce them to another mode of mathematical thinking.

Instruction should be highly exploratory and based on numerical and geometric experiences that capitalize on both calculator and computer technology. Instructional activities should be aimed at providing students with firm conceptual underpinnings of calculus rather than at developing manipulative techniques.

Discussion

The development of the calculus represents one of the great intellectual accomplishments in human history; perhaps the greatest achievement in the application of mathematics is the use of calculus in physics during the first third of this century. Today, methods of calculus are applied increasingly in the social and biological sciences, and in business as well. As students explore the topics proposed in this standard, it is important that they develop an awareness of, and appreciation for, the historical origins and the cultural contributions of the calculus.

The topics proposed for investigation in this standard should be developed as natural extensions of ideas students have previously encountered. The study of finite sequences and series recommended in the discrete-mathematics standard leads naturally to consideration of the corresponding infinite cases and concepts associated with limiting processes. Considerations of infinite sequences and series can occur at many different levels of abstraction and formalism. Consider, for example, the series

$$\frac{1}{2} + \frac{1}{4} + \frac{1}{8} + \frac{1}{16} + \frac{1}{32} + \ldots$$

At a very concrete level, this series can be summed in parts, as shown in figure 13.1.

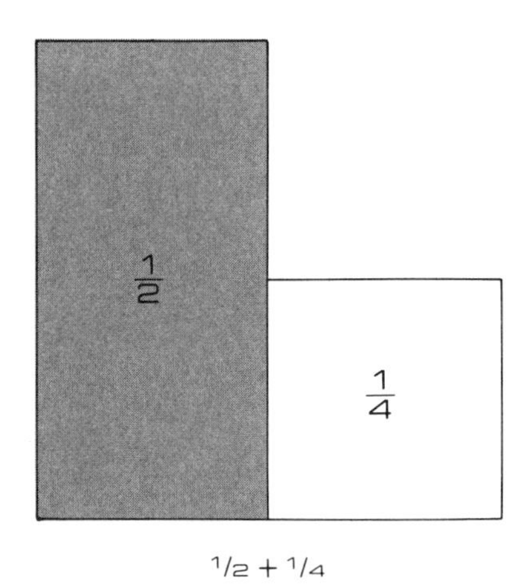

$^1/_2 + {}^1/_4$

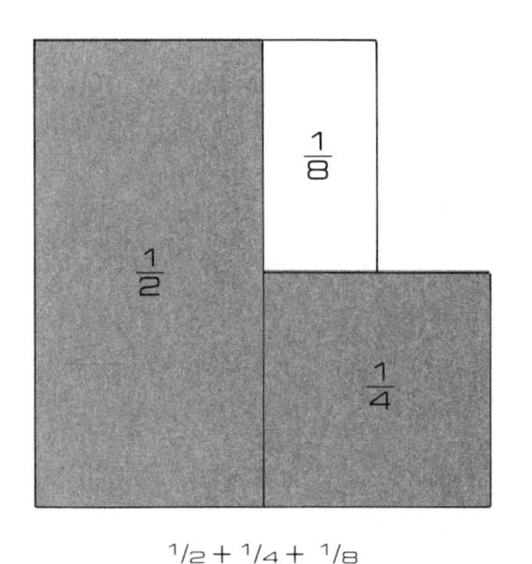

$^1/_2 + {}^1/_4 + {}^1/_8$ ...

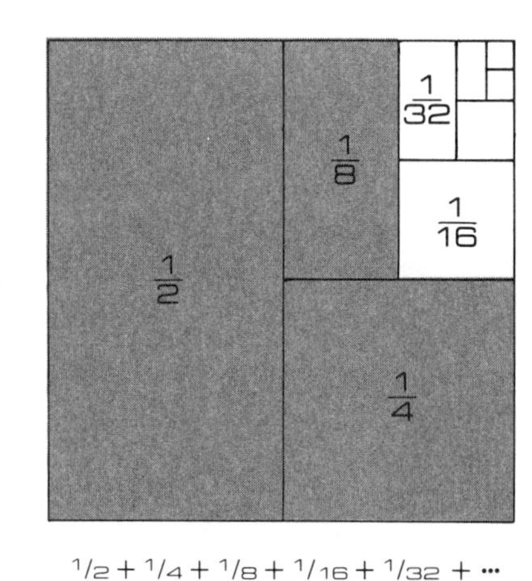

$^1/_2 + {}^1/_4 + {}^1/_8 + {}^1/_{16} + {}^1/_{32} + \cdots$

***Fig. 13.1.* An informal approach to infinite series**

At a somewhat higher but still intuitive level, the series can be summed by investigating the limit of the sequence of partial sums (either by using a calculator or a simple looping computer program or by applying the formula for the sum of a finite geometric series). An understanding of the concept of limit in contexts such as this should in turn contribute to the meaningful development of the remaining topics in this standard.

Similarly, it is important that all students recognize how the concept of the area under a curve builds on and extends their previous experiences with areas of geometric figures. College-intending students also should recognize how the rate of change builds on and extends their experiences with uniform motion and associated rates in algebra and trigonometry, and how the slope of a tangent to a curve generalizes the notion of the slope of a line as developed in algebra.

Many of these concepts can be approached through informal activities that focus on the understanding of interrelationships. For example, the concepts of slope, derivative, velocity, and acceleration could be addressed through experiences such as the following:

Given the velocity-time graph shown in figure 13.2, construct reasonable distance-time and acceleration-time graphs for the same system.

Possible solutions are shown in figure 13.3.

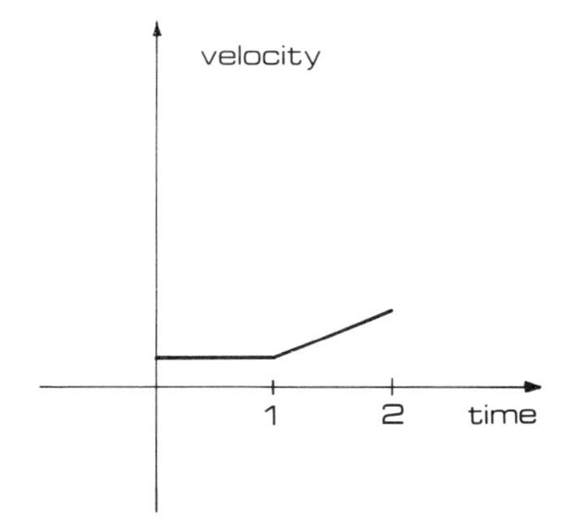

***Fig. 13.2.* A velocity-time graph**

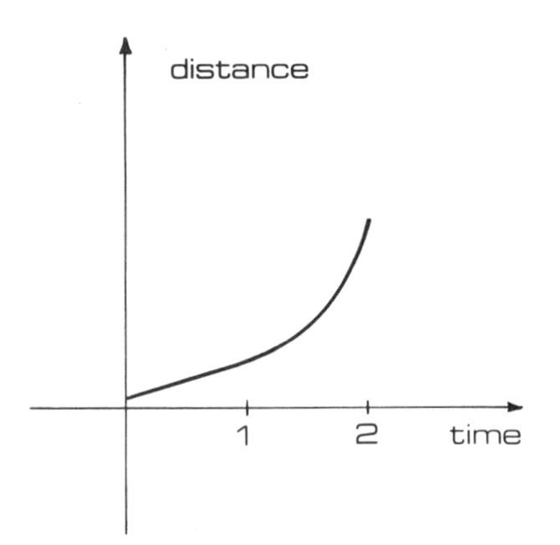

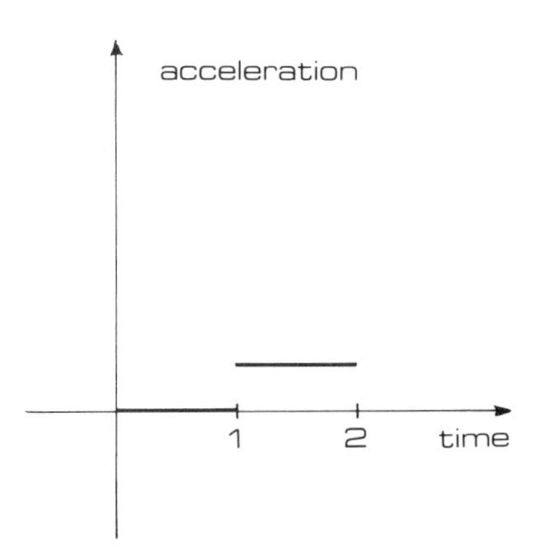

***Fig. 13.3.* Related distance-time and acceleration-time graphs**

Note that correct responses may vary; for example, the distance-time graph could be translated vertically.

Computing technology makes the fundamental concepts and applications of calculus accessible to all students. The area under a finite portion of a curve, for example, can be approximated geometrically by partitioning the region into rectangles with bases of equal length and heights given by function values at an endpoint of each base. Using a calculator or a computer-based algorithm, students can easily sum the areas to obtain a numerical approximation of the desired area. This approximation can then be sharpened by using sequences of rectangles whose bases are made to decrease toward zero. Project work might require students to investigate other ways of partitioning the region so as to obtain a more precise estimate. (A description of a process using trapezoids can be found in the standard on mathematical connections.)

All students could use a graphing utility to investigate and solve optimization problems, including the maximum-minimum problems traditionally associated with the first college-level course in calculus, without computing a derivative. (For an example, see Examples of Content Differentiation, pp. 134–36.) A great deal of mathematical understanding is reinforced in the context of solving these problems: data analysis, problem formulation, mathematical modeling, geometric topics, translation across multiple representations, and validation.

Using interactive graphing utilities, college-intending students could examine other characteristics of the graphs of functions, including continuity, asymptotes, end behavior (i.e., behavior as $|x| \to \infty$), and concavity. Moreover, such analysis can be applied with equal ease to the graphs of a variety of complicated functions and to curves specified by polar and parametric equations.

Computing technology also permits the foreshadowing of analytic ideas for college-intending students. From a computer-graphics perspective, for example, a differentiable function can be viewed as a function having the property that a small portion of its graph, when highly magnified, approximates a line segment. The "zoom in" feature permits students to magnify the graph of a function over a small interval to view the approximate segment representation and to compute the gradient (slope). Using the computer to plot successive gradients for a series of short intervals (see fig. 13.4) suggests a process that derives functions from functions.

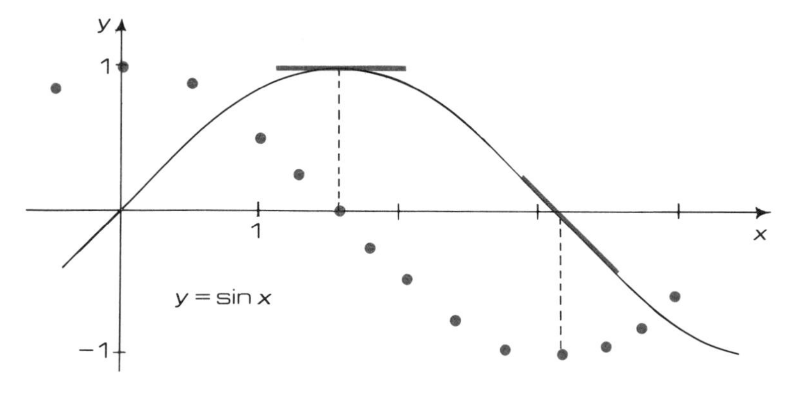

***Fig. 13.4.* Using computer graphics to see that the derivative of $y = \sin x$ is $y = \cos x$**

Although developing a foundation for the future study of calculus remains a goal of the 9–12 curriculum for college-intending students, equally important is the development of prerequisite understandings for further study of statistics, probability, and discrete mathematics. This standard calls for a new balance of skills, concepts, and applications in that por-

♦ ♦ ♦ ♦ ♦ ♦ ♦ ♦

tion of the curriculum traditionally associated with preparation for calculus. Instead of devoting large blocks of time to developing a mastery of paper-and-pencil manipulative skills, more time and effort should be spent on developing a conceptual understanding of key ideas and their applications. All students should have the benefit of a computer-enhanced introduction to some of the types of problems for which calculus was developed.

STANDARD 14: MATHEMATICAL STRUCTURE

In grades 9–12, the mathematics curriculum should include the study of mathematical structure so that all students can—

- ***compare and contrast the real number system and its various subsystems with regard to their structural characteristics;***
- ***understand the logic of algebraic procedures;***
- ***appreciate that seemingly different mathematical systems may be essentially the same;***

and so that, in addition, college-intending students can—

- ***develop the complex number system and demonstrate facility with its operations;***
- ***prove elementary theorems within various mathematical structures, such as groups and fields;***
- ***develop an understanding of the nature and purpose of axiomatic systems.***

Focus

The structure of mathematics is like the steel framework of a modern building. Students should become aware of this structure, how it provides a strong foundation on which a variety of content strands are built, and how it simultaneously holds these different strands together. For example, one of the girders in this building is the associative property to which are attached objects and operations in such wide-ranging mathematical subjects as arithmetic, algebra, functions, and geometric transformations. An awareness of these broad structuring principles frees students to take a more constructive approach to new mathematical topics and provides them with a conceptual framework that facilitates long-term retention.

Mathematical structure in the form of lists of general properties is *not* a good starting point for instruction. Rather, students gain a sense of the structure of mathematics over an extended time period through the general accumulation of experience, as well as through more focused activities. It is neither necessary nor appropriate for them to hear constantly the word *structure* applied to their activities; occasional summary statements will serve them far better. It also is essential to recognize that mathematical structure and formalism are not synonymous. In mathematics, just as with a building, all students can develop an understanding and appreciation of its underlying structure independent of a knowledge of the corresponding technical vocabulary and symbolism. The degree of formalism must be consistent with the student's level of mathematical maturity.

Discussion

Student insight into the structure of mathematics should be fostered at both micro and macro levels. At the micro level, it is gained through the pervasive application of the notion of step $(n + 1)$ following from step (n) because . . . ; for example, $3x = 7$ follows from $3x - 5 = 2$, since we know we can add 5 to each equation member without destroying the balance. Such arguments should be common to all mathematical study. At the macro level, insight is gained from the observation of quite different structures built on common features; the concept of an identity element, for example, plays a special role, not only in addition and multiplication within various number systems, but also in operations on functions, matrices, and geometric transformations.

Students come to understand the idea of structure through observation of the common properties in simple systems that seem on the surface to be quite dissimilar. For example, consider the structural commonality of binary multiplication, the logical operation "and," and a series electrical circuit (fig. 14.1).

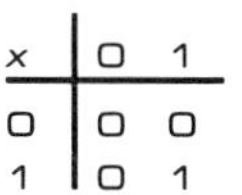

×	0	1
0	0	0
1	0	1

and	false	true
false	false	false
true	false	true

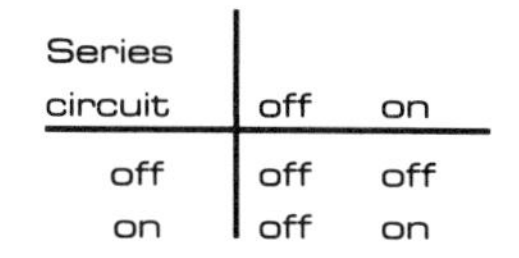

Series circuit	off	on
off	off	off
on	off	on

***Fig. 14.1.* Systems with structural commonalities**

The equivalent roles in these operation tables of 0, false, and "off," as well as of 1, true, and "on," are immediately evident.

The sets of symmetries of geometric figures, together with the operation of function composition, provide further concrete settings rich in opportunities for students to investigate mathematical structure. For example, the sets of symmetries for an isosceles triangle and a parallelogram each contain two elements, and their corresponding operation tables are structurally the same (fig. 14.2).

Students work with a series of number systems as they move through school. It is important that they return to this topic in order to view the associated structural characteristics of these systems. They should come to appreciate what is gained, what is lost, and what is retained in the structural characteristics of each new system. As students move from the integers to the rationals, for example, they retain addition and multiplication, they extend division, and they gain the density property. When college-intending students extend the real-number system to the complex numbers, they retain the field properties of the reals, they extend roots, but they lose order. For college-intending students, these ideas of structure should be extended to a more formal level through the organizing properties of groups and fields.

Applications of matrices as representation tools lead naturally to the usual definitions of equality and addition. All students could revisit this topic from a structural viewpoint. In particular, they could consider the set of 2×2 matrices with integer entries under addition and explore whether the commutative, associative, identity, and inverse properties hold. They could compare the resulting structure with other systems with which they are familiar (such as the integers under addition). College-intending students could continue a similar exploration with 2×2 real matrices using the usual definitions of matrix addition and multiplication. This would provide an example of a system in which multiplication is not commutative. In addition, these students could show that the existence of a multiplicative inverse for a given 2×2 matrix is equivalent to the existence of a unique solution of a related system of two linear equations in two variables.

Isosceles triangle

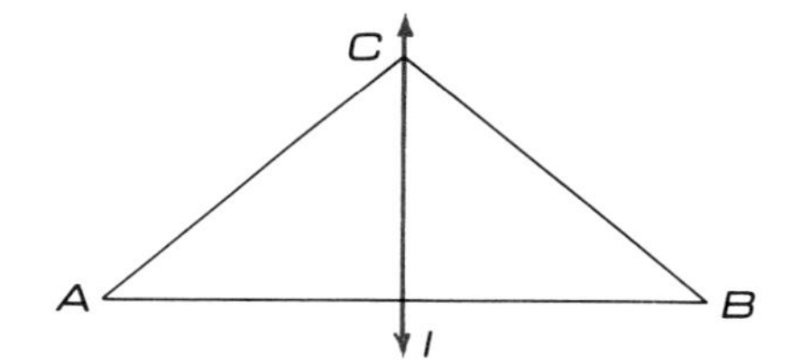

Reflection across line l maps $\triangle ABC$ onto $\triangle BAC$

Parallelogram

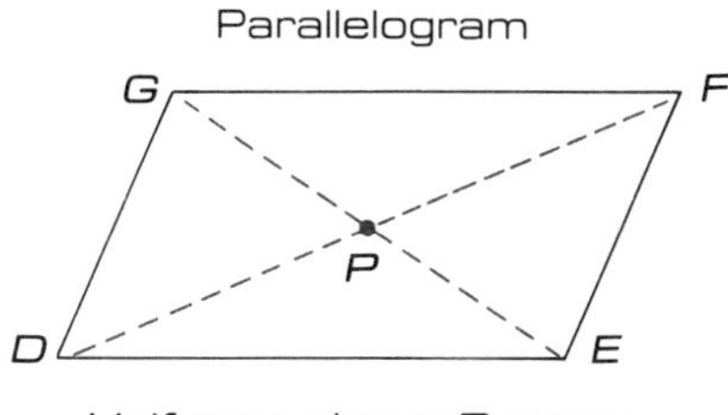

Half-turn about P maps $\square DEFG$ onto $\square FGDE$

Composition table

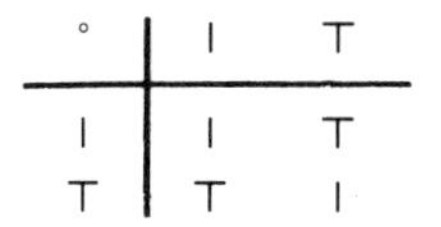

∘	I	T
I	I	T
T	T	I

I: Identity
T: Reflection across l or half-turn about P

***Fig. 14.2.* Sets of symmetries that are "essentially the same" under function composition**

College-intending students should also see how their school mathematics fits into the larger picture of advanced mathematical studies. For example, an important concept and a fundamental connecting link of mathematics is *isomorphism*; a direct translation of its Latin roots (*iso* = same, *morph* = form) conveys the term's essential meaning. Exponents (or, equivalently, logarithms) provide an excellent first example of an isomorphism that can reasonably be conveyed to students. In the world of numbers, we have multiplications, $b^m \times b^n$, but when this problem is translated into the world of exponents, the operation becomes addition, $m + n$. The result, translated back into the world of numbers, is b^{m+n}. In this way, students can gain an insight—through the lens of the fundamental property of exponents "when multiplying powers with the same base, add the exponents"—into a deep mathematical concept, often expressed in abstract form as $f(a * b) = f(a) \blacktriangle f(b)$. How this also applies to logarithms may be seen by direct substitution of log for f, $\times$ for $*$, and $+$ for $\blacktriangle$.

The concept of isomorphism also provides a means of clarifying some of the important connections between algebra and geometry. For example, college-intending students should recognize that the complex numbers under addition are isomorphic to two-dimensional translations under composition.

Consider the sum $(3 + i) + (-5 + 2i)$ and the isomorphism indicated in figure 14.3.

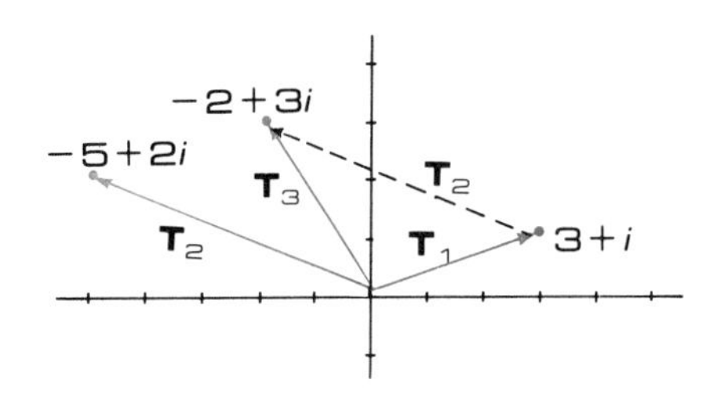

Complex number		Translation
$3+i$	corresponds to	$\mathbf{T}_1$: 3 units to the right, 1 unit up
$+$	corresponds to	composed with (followed by)
$-5+2i$	corresponds to	$\mathbf{T}_2$: 5 units to the left, 2 units up
$-2+3i$	corresponds to	$\mathbf{T}_3$: 2 units to the left, 3 units up

$$(3+i)+(-5+2i) \rightarrow \mathbf{T}_2 \circ \mathbf{T}_1 = \mathbf{T}_3 \rightarrow -2+3i$$

Fig. 14.3. Application of an isomorphism

As an enrichment activity, some students could be encouraged to establish that the nonzero complex numbers under multiplication are isomorphic to the spiral similarities (composites of a rotation and dilation with the same center) under composition.

Although the curriculum standards for grades 9–12 recommend that Euclidean geometry no longer be developed as a complete axiomatic system, students (especially the college intending) should nevertheless develop an understanding of the nature and purpose of axiomatic systems. This objective can best be achieved through cumulative experiences across several contexts. Students should develop short sequences of Euclidean geometry theorems in which they explore the role of assumptions and the consequences of replacing them by other statements that also hold in Euclidean geometry. They should investigate some of the simpler consequences of deleting or contradicting the Euclidean parallel axiom. Finally, they should have experiences with reasoning from assumptions in connection with their study of other topics in the curriculum, such as the structure of number systems and algebra. This shift in emphasis provides students with a better perspective on, and appreciation for, the role and nature of both geometry and axiomatic systems in contemporary mathematics.

Throughout the *Standards*, there has been a consistent emphasis on organizing the curriculum and instruction so that the connectedness of mathematical ideas is established and capitalized on both in problem solving and in the learning of new content. The study of mathematics in grades 9–12 should also provide students with an awareness and appreciation of the broad underlying themes and logical consistency of mathematics.

EVALUATION

♦ ♦ ♦ ♦ ♦ ♦ ♦ ♦

EVALUATION STANDARDS

OVERVIEW

This section presents fourteen evaluation standards, categorized by focus: General Assessment, Student Assessment, and Program Evaluation.

General Assessment

1. ***Alignment***
2. ***Multiple Sources of Information***
3. ***Appropriate Assessment Methods and Uses***

Student Assessment

4. ***Mathematical Power***
5. ***Problem Solving***
6. ***Communication***
7. ***Reasoning***
8. ***Mathematical Concepts***
9. ***Mathematical Procedures***
10. ***Mathematical Disposition***

Program Evaluation

11. ***Indicators for Program Evaluation***
12. ***Curriculum and Instructional Resources***
13. ***Instruction***
14. ***Evaluation Team***

Evaluation and Change

A common response to the challenge of the *Standards* is, "Yes, but who will change the tests?" Although pragmatic, this question shifts responsibility for change away from the individual to some unnamed higher authority. More productive—and more likely to make the vision embodied in the *Standards* a reality—are such responses as, "In what ways does the curriculum need to be changed?" "How best can these changes be made?" "How will we know when we have reached the *Standards*?" It is in the answers to these questions that the role of evaluation emerges as a critical component of reform. Evaluation is a tool for implementing the *Standards* and effecting change systematically. The main purpose of evaluation, as described in these standards, is to help teachers better understand what students know and make meaningful instructional decisions. The focus is on what happens in the classroom as students and teachers interact. Therefore, these evaluation standards call for changes beyond the mere modification of tests.

Yes, tests also need to change. They must change because a curriculum that fulfills the *Standards* will differ significantly, in both content and instruction, from most existing curricula. Many existing tests cannot measure the student outcomes identified in the *Standards*, for example, having K–4 students use many computation techniques and having 9–12 students make connections among mathematical topics. As the curriculum changes, so must the tests. Tests also must change because they are one way of communicating what is important for students to know. The tested curriculum can strongly influence what students are taught.

♦ ♦ ♦ ♦ ♦ ♦ ♦ ♦

In this way tests can effect change. Finally, existing tests must change because they are based on different views of what knowing and learning mathematics means. Knowing mathematics by doing mathematics in a technological world differs from developing a sequence of skills or objectives when calculators and computers did not exist and when mathematical applications were primarily confined to the physical sciences and commerce.

Evaluation is fundamental to the process of making the *Standards* a reality. Just as the Curriculum Standards propose changes in K–12 content and instruction, the Evaluation Standards propose changes in the processes and methods by which information is collected. These changes, developed in light of current knowledge about evaluation, are intended to increase and improve the gathering of relevant, useful information. Assessment and program-evaluation practices must change along with the curriculum. The Evaluation Standards propose that—

- student assessment be integral to instruction;
- multiple means of assessment methods be used;
- all aspects of mathematical knowledge and its connections be assessed; and
- instruction and curriculum be considered equally in judging the quality of a program.

Other Issues

When programs are evaluated and students are assessed, the information collected must be aggregated to draw meaning from what was observed or measured. The aggregated information is reported, used to assign grades, or used as an indication of the quality of the program. Although the process of aggregating information, scoring, and reporting is not discussed in detail in these Evaluation Standards, these issues are important and should be addressed. Because the Evaluation Standards call for multiple means of assessment, they imply that a variety of scoring schemes are to be used. Such schemes can include ratings of students' work that are based on its overall quality or on the inclusion of specific features or parts. Record keeping can be writing notes on cards or on a computer or maintaining a portfolio of the students' work. Whatever the scheme, the results should constitute an accurate and thorough indication of the mathematics that students know. Merely adding scores on written tests will not give a full picture of what students know. The challenge for teachers is to try different ways of grading, scoring, and reporting to determine the best ways to describe students' knowledge of mathematics as indicated in these *Standards*.

Format of the Evaluation Standards

The Evaluation Standards are organized in three sections: General Assessment, Student Assessment, and Program Evaluation. The general-assessment standards discuss principles relevant to any form of assessment and program evaluation. The student-assessment standards consider aspects of mathematical knowledge that should be assessed, as derived from the Curriculum Standards. The program-evaluation standards examine the assessment of the extent to which a mathematics program is consistent with the *Standards*.

The format of the Evaluation Standards differs from that of the Curriculum Standards. Each evaluation standard begins with a statement about

♦ ♦ ♦ ♦ ♦ ♦ ♦ ♦

the topics or concepts that an assessment or program evaluation should address, followed by a list of indicators describing what can be observed. Indicators are used to identify outcomes because they denote a measure placed in some context. In this situation, the Curriculum Standards serve as the criteria against which to compare the evidence of what students can do.

Emphases of the Evaluation Standards

The fourteen Evaluation Standards emphasize aspects of assessment and program evaluation that depart from current practice. The following aspects are to receive increased and decreased attention:

Increased Attention	Decreased Attention
♦ Assessing what students know and how they think about mathematics	♦ Assessing what students do not know
♦ Having assessment be an integral part of teaching	♦ Having assessment be simply counting correct answers on tests for the sole purpose of assigning grades
♦ Focusing on a broad range of mathematical tasks and taking a holistic view of mathematics	♦ Focusing on a large number of specific and isolated skills organized by a content-behavior matrix
♦ Developing problem situations that require the applications of a number of mathematical ideas	♦ Using exercises or word problems requiring only one or two skills
♦ Using multiple assessment techniques, including written, oral, and demonstration formats	♦ Using only written tests
♦ Using calculators, computers, and manipulatives in assessment	♦ Excluding calculators, computers, and manipulatives from the assessment process
♦ Evaluating the program by systematically collecting information on outcomes, curriculum, and instruction	♦ Evaluating the program only on the basis of test scores
♦ Using standardized achievement tests as only one of many indicators of program outcomes	♦ Using standardized achievement tests as the only indicator of program outcomes

GENERAL ASSESSMENT

The general-assessment standards present principles for judging assessment instruments. These principles are relevant for assessment at all levels and provide a rationale for the student-assessment standards that follow. These principles also apply to program evaluation. Inherent in the general-assessment standards is an assumption that all evaluation processes should use multiple assessment techniques that are aligned with the curriculum and consider the purpose of an assessment.

The vision of mathematics education in the *Standards* places new demands on instruction and forces us to reassess the manner and methods by which we chart our students' progress. In an instructional environment that demands a deeper understanding of mathematics, testing instruments that call for only the identification of single correct responses no longer suffice. Instead, our instruments must reflect the scope and intent of our instructional program to have students solve problems, reason, and communicate. Furthermore, the instruments must enable the teacher to understand students' perceptions of mathematical ideas and processes and their ability to function in a mathematical context. At the same time, they must be sensitive enough to help teachers identify individual areas of difficulty in order to improve instruction.

Many assessment techniques are available, including multiple-choice, short-answer, discussion, or open-ended questions; structured or open-ended interviews; homework; projects; journals; essays; dramatizations; and class presentations. Among these techniques are those appropriate for students working in whole-class settings, in small groups, or individually. The mode of assessment can be written, oral, or computer-oriented.

These and other techniques reflect the diversity of instructional methods implied by the Curriculum Standards and the various ways in which students learn, allow for diversity in student responses and modes of processing information, and provide reliable and valid information. Instructional decisions should be based on the convergence of information from different sources that supports or corroborates the need for a given educational response. When available information is contradictory, as for example when a student achieves good test scores but is unable to communicate mathematical processes, an assessment must search for deeper explanation. Simply put, assessment should not rely on a single instrument or technique.

STANDARD 1: ALIGNMENT

Methods and tasks for assessing students' learning should be aligned with the curriculum's—

- ***goals, objectives, and mathematical content;***
- ***relative emphases given to various topics and processes and their relationships;***
- ***instructional approaches and activities, including the use of calculators, computers, and manipulatives.***

Focus

The assessment of students' mathematics learning should enable educators to draw conclusions about their instructional needs, their progress in achieving the goals of the curriculum, and the effectiveness of a mathematics program. The degree to which meaningful inferences can be drawn from such an assessment depends on the degree to which the assessment methods and tasks are aligned or are in agreement with the curriculum. Little information is produced about students' mastery of curricular topics when the assessment methods and tasks do not reflect curricular goals, objectives, and content; the instructional emphases of the mathematics program; or how the material is taught.

When assessment instruments are aligned with the curriculum, the curriculum becomes the standard against which an assessment instrument should be judged. This alignment can be determined by examining the extent to which the instruments measure the content of the curriculum; are consistent with its instructional approaches, particularly the use of calculators, computers, and manipulatives; and cover the range of topics weighted according to the emphases of the curriculum.

Discussion

Alignment is a critical issue in the development or selection of assessment instruments or in the use of assessment data. Teachers, test developers, and administrators all must be concerned about the alignment of curricula and assessment, although their interpretations of assessment and its scope may vary. For teachers, *curriculum* can refer to the material covered in a chapter, unit, semester, or year; for administrators, it can be the content of the entire mathematics program; for test developers, it can refer to content that is common to the instructional programs of a number of school districts. Regardless of perspective, the same considerations are essential in determining the alignment of an assessment instrument with a curriculum.

The following discussion examines alignment with respect to the curricular aspects specified in these standards. Although the discussion focuses on assessments that are broad in scope, such as end-of-year or standardized tests, the issues are relevant to classroom assessment as well.

Goals, Objectives, and Mathematical Content. For assessment to be properly aligned, the set of tasks on the assessment instrument must reflect the goals, objectives, and breadth of topics specified in the curriculum. Ideally, all topics in the curriculum should be assessed. Assessment instruments aligned with the secondary school core curriculum, for example, will include tasks reflective of all the 9–12 standards.

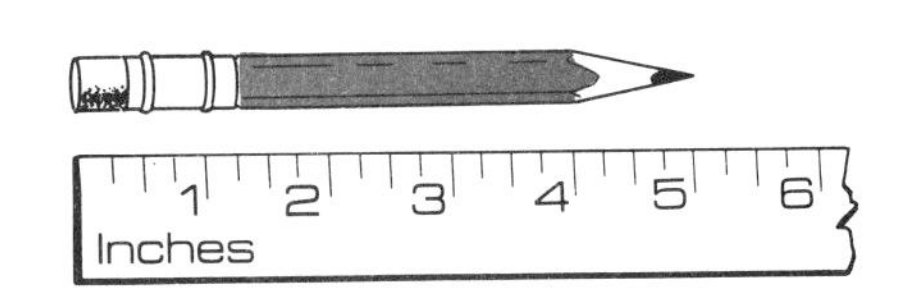

How long is the pencil?

***Fig. 1.1.* A poor task for assessing measurement**

To determine the alignment of an instrument with a curriculum, simply comparing the objectives and content of an assessment instrument with the objectives and content of the curriculum is insufficient. Individual items must be examined to determine the degree to which they measure the mathematical content that they purport to measure; that is, whether they are aligned in terms of content. For example, a student's capability of measuring length or distance using an appropriate instrument is not adequately assessed by items such as the one in figure 1.1.

To assess this skill, information is needed about whether each student can select an appropriate measurement tool, use it correctly (i.e., align and iterate it if necessary), and read the result. This information is best obtained through tasks that require a student to think about what mathematics is needed, to select a measuring tool, and to make an actual measurement. Assessment can be based on the student's answers or on observations of actions in the task. Only in this manner can a student's growth in mathematical power be determined.

The format of an assessment is an important factor that should be considered in determining the alignment of items and content. Other areas that might require particular formats for assessment include communication (which may involve talking, listening, or writing), reasoning (which might involve justifying or explaining responses), problem solving (which might involve recording processes as well as results), and estimation.

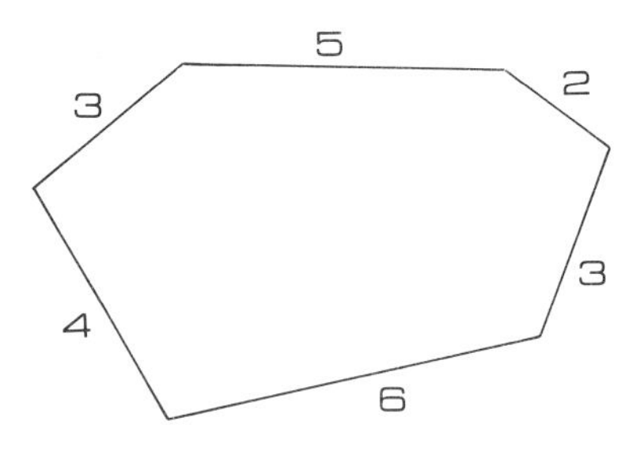

Find the perimeter.

***Fig. 1.2.* A poor task for assessing students' knowledge of perimeter**

Another factor that affects the alignment of items and content is whether students answer items correctly for the wrong reasons. A common example is given in figure 1.2.

Students may correctly add the lengths of the sides because that is the most obvious way to use all the given information. Little or no knowledge of perimeter is required. An item that will give a better indication of students' knowledge of perimeter is this:

Draw a six-sided irregular polygon with a perimeter of 23 units. Show all dimensions.

Items should be judged against the mathematics *as described in the curriculum.* Thus, one must be alert to differences in the interpretation of topics in an assessment instrument and in a curriculum. For example, problem-solving items can be anything from simple one-step story problems to open-ended, multistep problems. A simple story problem can measure problem solving according to the criteria established by the developers of an assessment instrument yet not measure problem solving as defined in the curriculum. The interpretation of content is of particular importance in selecting an assessment instrument that has been constructed by others, such as a standardized test.

Relative Emphases, Processes, and Relationships. The degree of alignment also depends on the extent to which the assessment's relative emphases on various topics and processes reflect the curricular emphases. An assessment instrument that contains many computational items and relatively few problem-solving questions, for example, is poorly aligned with a curriculum that stresses problem solving and reasoning. Similarly, an assessment instrument highly aligned with a curriculum that emphasizes the integration of mathematical knowledge must contain tasks that require such integration. And, for a curriculum that stresses mathematical power, assessment must include tasks with nonunique solutions.

Instructional Approaches and Activities. A final consideration is the extent to which the assessment reflects important aspects of instruction, such as the use of manipulatives, calculators, and computers. When these materials are used during instruction, they should be available during assessment, as long as their use is consistent with the purpose of the assessment. For example, if students routinely use calculators for solving problems in class, they should also be able to use calculators during assessments of their problem-solving abilities. Similarly, if students' understandings are closely related to the use of physical materials, they should be allowed to use these materials to demonstrate their knowledge during an assessment.

To select an appropriate assessment instrument, one must consider more than whether calculators, computers, or manipulatives are permitted to be used. Test items must be appropriate for use with these materials. For example, a multiple-choice algebra test that gives all answers in radical form, for example, $(2 + \sqrt{3})/4)$, might not be as appropriate for students using calculators as a test that includes answers that are decimal approximations.

STANDARD 2: MULTIPLE SOURCES OF INFORMATION

Decisions concerning students' learning should be made on the basis of a convergence of information obtained from a variety of sources. These sources should encompass tasks that—

- ***demand different kinds of mathematical thinking;***
- ***present the same mathematical concept or procedure in different contexts, formats, and problem situations.***

Focus

The quality of judgments about students' knowledge depends on the consistency of the results obtained as well as on the alignment between the assessment tasks and the knowledge they purport to measure (see Evaluation Standard 1). Although written tests structured around a single correct answer can be reliable measures of performance, they offer little evidence of the kinds of thinking and understanding advocated in the Curriculum Standards.

When teachers find that students perform in consistent ways on various formatted tasks that demand a range of mathematical thinking or represent different aspects of mathematical thought, teachers can have confidence in the accuracy of their judgments. As defined here, mathematical knowledge reflects the integration of mathematical concepts and procedures in a coherent, meaningful structure.

To achieve this integration, the *Standards* advocate an inquiry approach to instruction, stressing problem solving as the medium through which mathematics is learned and applied. Assessment, in its turn, is conceived of as an integral element of instruction, a method of generating valuable information for both student and teacher. As instruction explores content through an array of problem-solving situations, so too must assessment. In addition, to achieve the full range of goals articulated in the *Standards,* a variety of assessment formats is necessary; they should include written and oral tests, observations, essays, and performance evaluations.

Discussion

Within the instructional context, teachers continually make informal judgments about their students' progress. Nonetheless, they often are reluctant to use these observations as the basis of important instructional decisions because of their potential subjectivity and unreliability. If we are to assess the types of mathematical thoughts and actions delineated in the *Standards,* the richer information that results from a variety of assessment methods is not only desirable but essential. For example, students who can easily find the area of rectangles when given linear measurements often are puzzled when presented with a grid and asked to find the area of an irregular plane figure. In such a situation, we might ask, How well do they understand the notion of area as a covering? How do they view their actions when they multiply the length and width?

A constant theme in the Evaluation Standards is the need for multiple sources of information. The tasks described here cover many aspects of a single concept, use procedures in many contexts, and require students

♦ ♦ ♦ ♦ ♦ ♦ ♦ ♦

to integrate knowledge, particularly through problem-solving activities. These types of assessment are intrinsically valuable because they extend the experiences of the student and offer important instructional feedback to the teacher. But they have an extrinsic value as well because they allow teachers to develop a more complete picture of students' performance and mathematical power to give to both students and their parents.

When a student performs similarly on many tasks, teachers can have confidence in their judgment of that student. Discrepancies in performance can also be useful because they can identify difficulties that might go undiscovered if an assessment is made with a single instrument. For example, a student might perform well on individual written assignments but be unwilling or unable to describe his or her problem-solving approach during group discussions. Another student might be able to apply rules in a familiar context but fail to recognize the appropriate procedure when the task is placed in an unfamiliar context. Such discrepancies should suggest other questions: Is the behavior of the first student the result of shyness, lack of confidence, or insufficient understanding? Does the second student experience difficulty with understanding the task or the underlying procedure? How can each student be helped?

The advantage of using several kinds of assessments, some of which are embedded in instruction, is that students' evolving understanding can be continuously monitored. The disadvantage is that such a procedure is perceived as cumbersome. Records of students' progress should be more than a set of numerical grades or checklists—they can include brief notes or samples of students' work. Such records are evidence of students' continued growth in understanding. Students should also maintain their own records. At all grades, students can keep portfolios of their work; in the higher grades, as they become more verbally fluent and reflective, they should be encouraged to keep a mathematics journal. These journals can contain goals, discoveries, thoughts, and observations, as well as descriptions of activities. Journals not only allow students to chart their progress in understanding but also act as a focus for discussion between student and teacher, thereby fostering communication about mathematics itself. For example, indications that a student understands rational numbers and the record to be kept for each such indication might include the following:

- Correctly answering test items on computing with fractions, percents, and decimals. (Record the percentage of correct items and note the range of topics tested.)

- Solving a homework problem situation by depicting graphically and in other forms how the speed of a car changes as it is driven through a city. (Record a score based on a four-point scale—four points for accurately drawing a graph that includes all the details of the change in speed, including stops, accelerations, and decelerations; three points if the general shape of the graph is given without the attention to detail; two points if the approach was correct with some errors in the details and shape of the graphs; one point if some notion of proportional thinking was evident but the graph does not represent the situation and details are missing; or no points for not trying.)

- Accurately representing, in a paper for civics class, the chances of winning a lottery and correctly using words like *odds, proportion,* and *probability.* (Student records in a log the mathematical concepts used in other subject areas.)

- Successfully using proportional scales and charts to format pages in print shop. (Write a short note of conversation with the printing teacher and put in the student's folder.)

- Being observed in class helping another student solve a percent problem. (Make a note of the use of percent and put in the student's folder.)

- Painting in graphics-art class an enlargement of a picture from a magazine. (File student's notes of what was done to ensure that proportions were correct.)

The learning of mathematics is a cumulative process that occurs as experiences contribute to understanding. A numerical score or grade assigned at a single point in time offers only a glimpse of students' knowledge. If the goal of assessment is a valid and reliable picture of students' understanding and achievement, evidence must come from a variety of sources. The assessment process described in this standard will produce a more complete and valid indicator of achievement than that possible from a single type of instrument.

♦ ♦ ♦ ♦ ♦ ♦ ♦ ♦

STANDARD 3: APPROPRIATE ASSESSMENT METHODS AND USES

Assessment methods and instruments should be selected on the basis of—

- ***the type of information sought;***
- ***the use to which the information will be put;***
- ***the developmental level and maturity of the student.***

The use of assessment data for purposes other than those intended is inappropriate.

Focus

The purpose of an assessment—to identify areas of difficulty for individual students, to gather data for instructional planning, to assign grades, or to evaluate a program—should dictate the kinds of questions asked, the methods employed, and the uses of the resulting information. When one type of measure is used in lieu of another, the information obtained is often invalid or useless. In addition, the methods used to gather information should be appropriate to the developmental level and maturity of the students. Table 3.1 outlines some common purposes of student evaluation. The purposes in the table overlap, and the questions are only examples.

Discussion

The assessment of student performance serves many purposes. For the student, assessment aids learning and measures mathematical knowledge and power. For the teacher, it provides information about how instruction should be modified and paced. For the administrator, it charts the effectiveness of a program. In addition, the general public expresses concern about academic achievement. Each of these groups asks different questions. Each needs different kinds of information. An assessment designed to answer one kind of question can misrepresent the answer to another. Although this caveat might seem obvious, the legitimate uses of an assessment often are not well understood or defined, as when, for example, standardized tests of general mathematical achievement are used for curriculum evaluation.

The information required by different audiences lies on a continuum from the most specific, such as measures of a curriculum objective, to indicators of general mathematics performance. Teachers who want to know how they can help Sandra or John understand fractions as parts of a region, for example, will learn best by questioning students about their perceptions of a specific concept. However, when the same teachers ask how well Sandra or John has understood the course material as a whole, they must cast a wider net; assessment methods limited to a single aspect of learning will not suffice. Students should have the opportunity to show how well they have integrated their knowledge by applying their learning in a larger context.

TABLE 3.1
PURPOSES AND METHODS OF ASSESSMENT

Purpose (examples of questions asked)	For Whose Use	Unit of Assessment	Type of Assessment	Assessment Methods
Diagnostic ♦ What does this student understand about the concept or procedure? ♦ What aspects of problem solving are causing difficulty? ♦ What accounts for this student's unwillingness to attempt new problems or see the application of previously learned materials?	Individual teacher Individual student	Individual student	♦ Tasks that focus on a specific skill, type of procedure, concept, strategy, or a type of reasoning ♦ Each student evaluated	♦ Observation ♦ Oral questions that ask students to explain their procedures ♦ Focused written tasks ♦ Directed test items
Instructional Feedback ♦ What do students know about the material presented? ♦ Can students apply their learning to new situations? ♦ Do students understand the connections among ideas? ♦ How shall I pace instruction? ♦ Does the class need more intensive review or more challenging material?	Individual teacher	Class	♦ Tasks that require an integration of knowledge ♦ Tasks that cover a range of skills, concepts, and procedures ♦ Tasks that require the application of learning to new contexts ♦ Problem-solving and reasoning tasks ♦ Tasks that vary the format and context in which the material is presented ♦ Matrix-sampling test situations	♦ Written tests, including those that require differential methods for solutions to problems ♦ Class presentations ♦ Extended problem-solving projects ♦ Observation of class discussion ♦ Take-home tests ♦ Homework, journals ♦ Group work and projects
Grading ♦ How well has this student understood and integrated the material? ♦ Can this student apply his or her learning in other contexts? ♦ How prepared is this student to proceed to the next grade or level?	Individual student Parents School	Individual student	♦ Tasks that demand the integration of material that was taught ♦ Tasks that are intrinsically interesting and challenging to the student ♦ Tasks that require the student to structure the material and generate solutions, in the context of the real world, as well as in mathematics	♦ Extended problem-solving projects ♦ Papers or written arguments that demand thoughtful inquiry about a mathematical topic ♦ Written tests that present problems with a range of difficulty based on expectations for course ♦ Oral presentations

♦ ♦ ♦ ♦ ♦ ♦ ♦ ♦

TABLE 3.1—CONTINUED
PURPOSES AND METHODS OF ASSESSMENT

Purpose (examples of questions asked)	For Whose Use	Unit of Assessment	Type of Assessment	Assessment Methods
Generalized mathematical achievement ♦ How does the general mathematical capability of this student compare with others or with a national norm?	Parents Teachers Administrators	Individual student	♦ Tasks organized in highly reliable tests designed for maximum discrimination among students	♦ Standardized achievement tests
Program Evaluation ♦ How effective is this instructional program in achieving our goals for mathematical learning?	Teachers Administrators Other decision makers	Class School	♦ Tasks that reflect the intent of the curriculum goals ♦ Tasks that are aligned to the instructional methods and content of the curriculum (see Standards 12 and 13) ♦ Matrix-sampling test situations	♦ Student interviews ♦ Performance tests ♦ Criterion-referenced tests ♦ Observation of class discussions ♦ Success of students who have completed the program

When assessments are used for grading or as summative indicators of achievement, a number of measures should be used (see Evaluation Standard 2). In addition to the more traditional types of written tests, students should be given tasks that are challenging and complex and that allow them to perform at their maximum level of ability. Not only should such tasks provide meaningful information to parents and school authorities, but they should be interesting and valuable experiences in their own right, giving the student a sense of accomplishment.

Criterion-referenced tests, often used for program evaluation, can cover a narrower range of content. Generally designed to measure the attainment of specific objectives, they are, in some sense, a collection of minitests that each focus on a relatively narrow area of achievement. When the objectives on these tests reflect the goals of the school or the program, the tests are a valid measure of effectiveness. However, such tests usually present material in only one format (written multiple choice), which limits what can be measured. Thus, other information is needed to confirm their results.

In contrast to assessments that yield useful curriculum-based information to the teacher and student, more general types of tests are relatively insensitive to individual curricula. The standardized achievement test is the most prominent example. Its purpose is to measure an individual student's relative position in a population. Consequently, this type of test must maximize individual differences among students while measuring the common elements of their instruction. As a result, it is less appropriate for measuring the effectiveness of any specific curriculum and is more likely to reflect a student's general achievement, background, and prior knowledge. Furthermore, because of their format, standardized norm-referenced tests have difficulty measuring the generation of ideas, the formulation of problems, and the flexibility to deal with mathematical problems that are not well structured (i.e., problems similar to those encountered in everyday life). For these reasons, they are inappropriate as the only measure of whether teaching and curricula reflect the spirit of the *Standards.*

Whatever the purpose of the assessment, the methods used should consider the characteristics of the students themselves. Students' mathematical and cognitive development is a gradual, cumulative process built on prior experience and understandings. This consideration is particularly important in the early years, when basic skills and concepts are initially encountered. For example, recording responses on op-scan sheets imposes an additional irrelevant demand on young children. The very results of written tests in the early grades can be suspect: when children use their developing reading and writing skills to process the mathematical content of questions, the results might be more reflective of achievement in these areas than of their understanding of mathematics. At this stage, when children's understanding is often closely tied to the use of physical materials, assessment tasks that allow them to use such materials are better indicators of learning.

Students differ in their perceptions and thinking styles. An assessment method that stresses only one kind of task or mode of response does not give an accurate indication of performance, nor does it allow students to show their individual capabilities. For example, a timed multiple-choice test that rewards the speedy recognition of a correct option can hamper the more thoughtful, reflective student, whereas unstructured problems can be difficult for students who have had little experience in exploring or generating ideas. An exclusive reliance on a single type of assessment can frustrate students, diminish their self-confidence, and make them feel anxious about, or antagonistic toward, mathematics.

Finally, problems that reflect the potential of some students can be enigmas to others, disqualifying them from making any kind of meaningful response. Prior knowledge, experience, and the opportunity to learn are important considerations in interpreting test results. A challenging task in problem solving for some students is an exercise in recall for others. In part, the solution to this dilemma lies in the types of tasks used in the assessment. The complex, multifaceted tasks advocated in these Evaluation Standards can be structured to allow students to answer at different levels of sophistication. If students are to perform at their maximum levels of ability, the measures by which they are judged should give them the opportunity and the encouragement to do so.

♦ ♦ ♦ ♦ ♦ ♦ ♦ ♦

STUDENT ASSESSMENT

If a conversation is to be meaningful to both of its participants, each must listen to the other; in the absence of such mutual attention, the conversation becomes an exercise in futility. Similarly, the act of teaching should be founded on dialogues between teachers and students, each responding to the other on the basis of what has been said or done. *Assessment* refers to the process of trying to understand what meanings students assign to the ideas being covered in these dialogues; as such, it is an integral element of effective teaching. Periodic assessment provides the teacher with a basis for deciding what questions should be asked and what examples and illustrations should be used; ultimately, it offers a foundation for any meaningful dialogue between teacher and student.

The student-assessment standards describe what is to be observed and measured in the process of understanding what mathematics students know. Teachers drawing meaning from their interactions with students is central to this process. At this level the most important decisions about student learning are made. The general-assessment standards offer principles for the student-assessment standards, but the latter are paramount for helping students acquire the knowledge of mathematics as described in the Curriculum Standards. Thus, the student-assessment standards are more specific and relevant to teachers.

Assessment must be more than testing; it must be a continuous, dynamic, and often informal process. It manifests itself in teachers' statements, for example, "Johnny seems to have difficulty in graphing equations" or "Wilma demonstrated a lot of insight in solving those addition and subtraction problems." Assessment is more than the establishment of definitive conclusions. Assessment is cyclic in nature, a process of observation, conjecture, and constant reformulation of judgments about students' understanding. Assessment should produce judgments that are evolutionary in nature, regardless of whether those judgments are based on classroom discourse or on the more formal aspects of testing that characterize nearly every instructional program.

Testing to assign grades is one of the most common forms of evaluation. But assessment is a much broader and basic task, one designed to determine what students know and how they think about mathematics. Assessment should produce a "biography" of students' learning, a basis for improving the quality of instruction. Indeed, assessment has no raison d'être unless it improves instruction.

Teaching is effective to the degree that it takes student thinking into consideration. Without ongoing communication, a teacher's instructional strategies can only randomly enhance learning. Consider the student who answers the following item incorrectly:

If the fractions represented by points *C* and *D* on this number line are multiplied, what point will best represent the product?

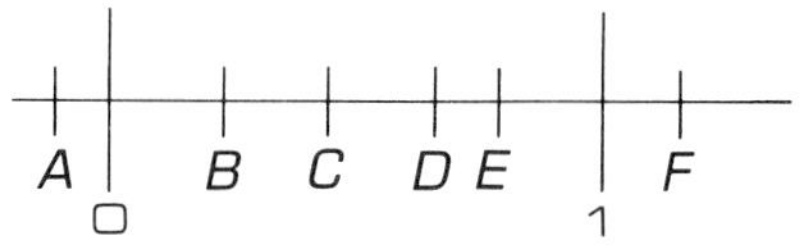

A B C D E F
0 1

If the student answers "point *A*," is he or she confusing multiplication with subtraction? Similarly, if points *E* or *F* are selected, is multiplication being confused with addition? Or is the answer a manifestation of the general misbegotten rule "multiplication makes bigger"? Did the student assign values to *C* and *D* but incorrectly multiply the fractions? What representation or model of multiplication is the student using? Would the student identify the same point if presented with the product $D \times C$ instead of $C \times D$? How would the student respond if specific fractions were assigned to C and D? Only through explicit and careful assessment of *how* a student does mathematics can instruction be tailored to individual needs, thereby enhancing a student's chances for success.

These seven student-assessment standards focus on assessing students' understanding of, and disposition toward, mathematics. Each presents a list of mathematical outcomes derived from the standard's central focus. These standards and their associated tasks can serve as heuristics for designing assessment systems that include instruments, procedures for aggregating data, and ways of record keeping that are comprehensive and that tap more than superficial understanding. It is reasonable—indeed, expected—that every lesson cover more than one aspect and that over a series of lessons most, if not all, aspects are addressed in both instruction and assessment. For example, whereas instruction in any particular procedure may not encompass all seven aspects of procedural knowledge, each of the seven aspects will be represented in both instruction and assessment over a series of lessons.

STANDARD 4: MATHEMATICAL POWER

The assessment of students' mathematical knowledge should yield information about their—

- ***ability to apply their knowledge to solve problems within mathematics and in other disciplines;***
- ***ability to use mathematical language to communicate ideas;***
- ***ability to reason and analyze;***
- ***knowledge and understanding of concepts and procedures;***
- ***disposition toward mathematics;***
- ***understanding of the nature of mathematics;***
- ***integration of these aspects of mathematical knowledge.***

Focus

In mathematics, as in any field, knowledge consists of information plus know-how. Know-how in mathematics that leads to mathematical power requires the ability to use information to reason and think creatively and to formulate, solve, and reflect critically on problems. The assessment of students' mathematical power goes beyond measuring how much information they possess to include the extent of their ability and willingness to use, apply, and communicate that information. The assessment should examine the extent to which students have integrated and made sense of information, whether they can apply it to situations that require reasoning and creative thinking, and whether they can use mathematics to communicate their ideas. Additionally, assessment should examine students' disposition toward mathematics, in particular their confidence in doing mathematics and the extent to which they value mathematics.

An assessment of students' mathematical power is broad in scope and should include *all* the aspects identified in this standard and determine the extent to which they are integrated. The assessment of mathematical power should not be construed as the assessment of separate or isolated competencies. Although one aspect of mathematical knowledge might be emphasized more than another in a particular assessment, it should remain clear that mathematical power concerns all aspects of mathematical knowledge and their integration.

Discussion

To have assessment practices in mathematics that reflect these standards, all important aspects of mathematical knowledge must be assessed. Student-assessment standards 5–10 address each of these aspects and how they are interrelated and give guidelines for assessing them. This discussion focuses on determining the extent to which these aspects are integrated.

Problem situations from different areas of study offer a rich context in which to assess students' mathematical power. Problems derived from such situations typically require students to apply a variety of mathematical concepts and procedures and engage in some form of mathematical reasoning. Understanding these concepts and their interrelationships is essential to interpreting a situation and deriving an appropriate plan of action. Knowing what procedures are appropriate or necessary and how to execute them is essential to carrying out the plan successfully. Furthermore, these problems require students to use various forms of reasoning to arrive at a solution. Consider, for example, the following task derived from a social studies lesson on the commerce between North America and Hong Kong:

Pretend you are a pilot for a major airline-transport company. You have been assigned for the first time to a trans-Pacific flight from New York to Hong Kong. You are curious about the shortest route between the two cities, but all you have is a regular globe and a piece of string. You know that the distance around the Earth along the equator is 25 000 miles. With only these two items, how can you figure the shortest distance? What is this distance?

The task requires students to interpret the shortest distance between two points on a sphere, find a way of measuring that distance, and use proportional reasoning inventively and creatively. The task is suitable for junior high school students. Students can work individually or in small groups while the teacher observes their interactions. These observations can yield information about the students' ability to apply their knowledge in solving the problem. This task illustrates how mathematics can be integrated into other areas of the school curriculum.

The assessment of students' mathematical power is appropriate at all grade levels and should not be delayed on the grounds that students must know a great deal of mathematics before they can integrate this knowledge. Group tasks are particularly useful in the lower grades for assessing the integration of students' mathematical knowledge. At the K–4 level, a group task like the following can be devised:

Materials Required

- **Large box of raisins**
- **Containers of different sizes**
- **Balance**
- **Calculator**

Task

1. **Estimate the number of raisins in the box.**
2. **Use any of the materials to make a better estimate.**
3. **Check your estimate by different methods.**
4. **Record your results and give an oral account of your work.**

The assessment of students' performance on this task focuses on their choice of an overall solution strategy. A successful approach entails counting the number of raisins in a small unit and relating the smaller unit to the whole (the large box) in some way. Students can use a balance to halve successively the number of raisins until a manageable number is obtained and then multiply to obtain the total in the large box. Or they can count the number of raisins in a small container and find the number of small containers necessary to fill the large box. Or they can measure the height of a small proportion of the raisins in the large box and the height of the full box and relate the two heights. Whatever their

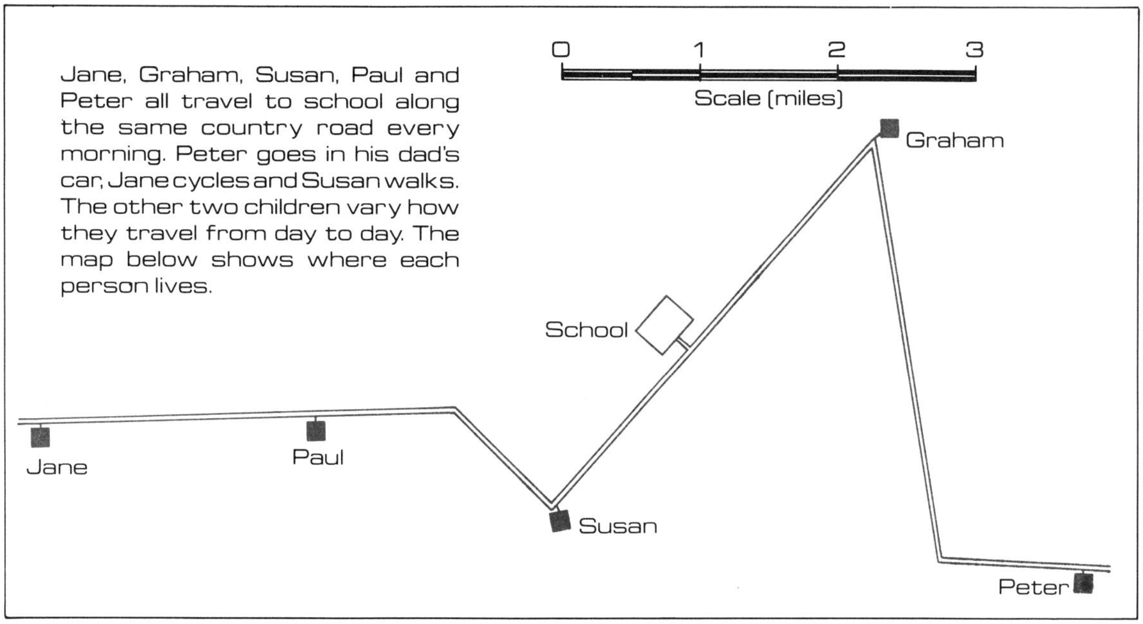

Jane, Graham, Susan, Paul and Peter all travel to school along the same country road every morning. Peter goes in his dad's car, Jane cycles and Susan walks. The other two children vary how they travel from day to day. The map below shows where each person lives.
0
1
2
3
Scale (miles)
Graham
School
Jane
Paul
Susan
Peter

Fig. 4.1. Map showing distance from school to home

strategy, students engage in counting, computing, measuring, and communicating to perform the tasks.

Reporting on the students' strategy can be done using a rating scale. Evidence of students' application of their strategy is obtained from the recorded results and oral accounts of what they did. Levels of mathematical argument can be similarly evaluated. An assessment of the mathematical concepts and procedures used can focus on the accuracy of counting and calculation; the correct choice of operation; the understanding and use of balance; and such calculator skills as accurate keying, recognition of the order of operations, and correct interpretation of the display. Finally, the assessment of communication skills can focus on the students' recorded and oral accounts of their work.

Problems and group tasks are not the only means of assessing the integration of students' mathematical knowledge. Written tasks can be used effectively. A written task can consist of multiple subtasks, encompassing various aspects of mathematical knowledge and their integration. Figures 4.1, 4.2, and 4.3 contain an example (Swan 1987, pp. 28–30) that illustrates this approach:

Length of journey to school (miles)
6
4
2
20
40
Time taken to travel to school (minutes)

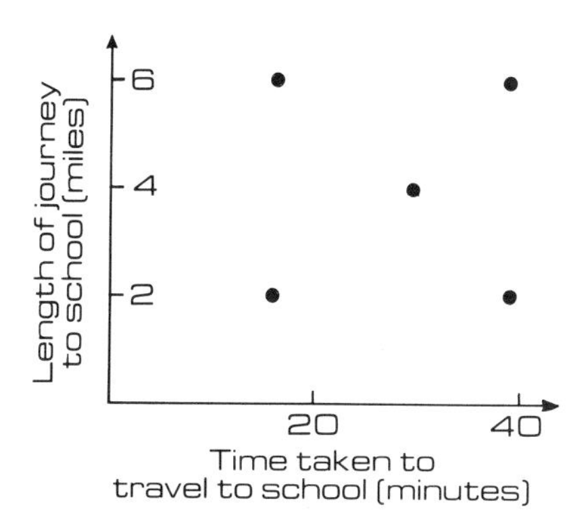

Fig. 4.2. Length and time of journey

The graph in figure 4.2 describes each pupil's journey to school last Monday.

a. **Label each point on the graph with the name of the person it represents.**

b. **How did Paul and Graham travel to school on Monday?**

c. **Describe how you arrived at your answer in part *b*.**

d. **Peter's father is able to drive at 30 mph on the straight sections of the road but has to slow down for the corners. Sketch a graph on the axes in figure 4.3 to show how the car's speed varies along the route.**

Peter's Journey to School
Car's speed (mph)
30
20
10
0
1
2
3
4
5
6
Distance from Peter's home (miles)

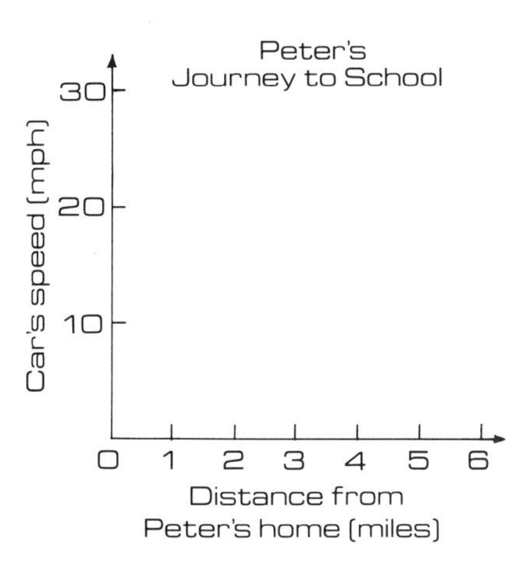

Fig. 4.3. Speed by distance for Peter's journey

The assessment of students' performance on this task focuses on their ability to apply several mathematical concepts, skills, and processes simultaneously. The task entails reading, interpreting, and selecting information from the map and combining it with information in the problem to estimate the children's relative travel times and establish the correspondences between points on the graph and the children. The selection and use of information necessary to label the points on the graph entails an understanding of the relationship among distance, time, and speed. Students must apply their knowledge of coordinate graphs and the variables represented to establish a correspondence between points and children. Additionally, they must infer from the graph the mode of travel used by two of the five children. The students' description of their reasoning in labeling those points offers evidence of their ability to communicate mathematical ideas. Finally, the task calls for the graphical representation of a car's speeds along the route depicted in the map, which requires the simultaneous consideration of the route, the additional information in part *d*, and the variables represented in the graph.

Students' responses to this task can be evaluated by rating their ability to read and interpret information in the map, combine this information with that in the problem statement, and translate and summarize it in a graph. Marks can be given for correctly identifying each point in figure 4.2. The students' description of their reasoning in locating those points can be assigned marks if Paul and Graham are identified as cycling or running or if the explanation matches the graph drawn in figure 4.2. The response to figure 4.3 can be assigned marks on the basis of the inclusion of the main features of the graph—correctly locating endpoints, indicating two minima in the correct position, showing the speed as 30 mph for at least 1 mile in the middle section of the graph, and correctly recording all other features of the graph. The score for the tasks would be the sum of the marks given.

Written tasks such as these can be useful in determining the extent to which students' knowledge of mathematics has been integrated. However, such assessment cannot be based solely on students' performance on a single task, regardless of how valid or appropriate the task might be. To assess the integration of mathematical knowledge, information must be obtained from several tasks performed over time in a variety of contexts.

♦ ♦ ♦ ♦ ♦ ♦ ♦ ♦

STANDARD 5: PROBLEM SOLVING

The assessment of students' ability to use mathematics in solving problems should provide evidence that they can—

- ***formulate problems;***
- ***apply a variety of strategies to solve problems;***
- ***solve problems;***
- ***verify and interpret results;***
- ***generalize solutions.***

Focus

If problem solving is to be the focus of school mathematics, it must also be the focus of assessment. Students' ability to solve problems develops over time as a result of extended instruction, opportunities to solve many kinds of problems, and encounters with real-world situations. Students' progress should be assessed systematically, deliberately, and continually to effectively influence students' confidence and ability to solve problems in various contexts. Giving students feedback about the results of this assessment, on both the processes used and the results attained, is critical to their development as problem solvers. In addition, an assessment can give helpful information to teachers regarding problem situations that are challenging, instructive, and interesting, and yet not defeating, for students.

Assessments should determine students' ability to perform all aspects of problem solving. Evidence about their ability to ask questions, use given information, and make conjectures is essential to determine if they can formulate problems. Assessments also should yield evidence on students' use of strategies and problem-solving techniques and on their ability to verify and interpret results. Finally, because the power of mathematics is derived, in part, from its generalizability (e.g., a two-space solution can be generalized to a three-space solution), this aspect of problem solving should be assessed as well.

Discussion

Methods for assessing students' ability to solve problems include observing students solving problems individually, in small groups, or in whole-class discussions; listening to students discuss their problem-solving processes; and analyzing tests, homework, journals, and essays. Feedback to students can have a variety of forms, including written or oral comments or numerical scores on a specific exercise. Scoring schemes include giving two scores (one for the answer and one for the strategies used); rating the student's work on a scale (e.g., 4—perfect, 3—nearly all correct with some computational errors, 2—right idea but poor execution, 1—tried the problem, 0—nothing was done); or giving points for such primary features as computation, pictures, tables, strategies, and verification. One way of reporting progress in problem solving is with a problem-solving profile. The profile can include ratings on a student's willingness to engage in problem solving, the use of a variety of strategies, facility in finding the solution to problems, and consistency in verifying the solution. The report to parents should include a sample of the student's problem-solving work.

The following are examples of situations appropriate for students at various grade levels that can be used to assess their abilities to solve problems successfully.

Grades K—4

1. *Formulate problems.*

You have this amount of change:

8 pennies

5 nickels

11 dimes

5 quarters

These items are for sale:

Box of cereal for $1.60

Glass of milk for $0.40

A poster for $0.90

One ball for $1.20

Use this information to make up a problem.

This situation can be given to students as a writing task. Students can be evaluated on how much of the given information they use, the reasonableness of their problem, and its mathematical sophistication. The problem "How many coins are there?" includes less of the given information than the problem "Do I have enough money to buy one of each item? If not, how much do I need?" The quality of the problem, however, is not necessarily dictated by the inclusion of all the information: For example, "If I buy a box of cereal, how many glasses of milk can I buy?" is a far more sophisticated problem but one that uses a minimum of given information. Students can work in small groups to generate problems. Calculators can be helpful.

2. *Solve problems.*

Read the following problem and answer the question posed:

Paula, Teresa, and Dale ran a race. Paula took three minutes and Dale took four minutes to finish the race. Who won the race?

This task can be used in instruction to determine if students recognize that essential information is missing. Once they determine that they need to know Teresa's time, other questions can be asked: Is it possible to give a time for Teresa so that she can win? Is it possible to give a time for Teresa so that Paula can win? It is important to observe if the information given is reasonable and if students can verbalize why a certain value for Teresa's time is given. This example shows how a routine exercise can serve as the basis for generating other tasks by deleting a condition, removing the question, or adding irrelevant information.

♦ ♦ ♦ ♦ ♦ ♦ ♦ ♦

3. *Apply strategies to solve problems.*

With a calculator, find three numbers whose product is 2431. Keep a record of what you do to find the answer.

This task is suitable for guessing-and-testing strategies and can be used in a testing situation. Students should be encouraged to write down their guesses and explain what they did. It should be observed if students take a systematic approach to developing guesses, if they keep a list of guesses, and if they place limitations on what the numbers can be (e.g., only odd numbers will work). The calculator is essential for generating a number of guesses in a short period of time. Students should be given points both for the right answers and for the use of one or more appropriate strategies. Random guessing should not be awarded strategy points. Some time should be available for students to explain their approaches.

Grades 5—8

1. *Solve problems.*

Students' pulse rates vary. What would be considered the normal pulse rate for students in your class? You might want to consider various characteristics or conditions (e.g., exercise) and find out how they relate to pulse rate.

The evaluation of this activity should focus on the reasonableness of students' questions, the various forms of representations used to report data, whether the results are verified, and whether any generalizations are made. This activity can extend over several days and is appropriate for small-group work. A score or rating for the total project can be given to each group. Specific parts of the students' work can be emphasized by scoring them separately, such as the number of representations used (table, graph, equations, and written report). A calculator and a computer are essential.

2. *Formulate problems; verify and interpret results.*

Four of every five dentists interviewed recommended Yukky Gum. Write a question to go with this statement to make a problem. Solve the problem.

This task can be embedded in instruction and is appropriate for large-group discussion. The assessment of students' performances on this task focuses on their choice of questions. The question must be logically connected to the statement. Consider, for example, the question, If 1000 dentists were interviewed, how many recommended Yukky Gum? Students' explanations will give some indication of how they viewed the task and what features they focused on in developing their questions. Requiring students to solve their own problems gives an indication of their ability to solve problems and encourages them to ask reasonable questions.

The activity can be extended by asking students to formulate a problem whose answer is "160 dentists." Writing a question to fit an answer requires an interpretation of the result in light of the conditions stated. The question should be judged on whether the result, 160 dentists, matches the given conditions. Some note of students' posing of appropriate questions can be recorded.

◆ ◆ ◆ ◆ ◆ ◆ ◆ ◆

3. *Apply strategies; solve problems; verify and interpret results.*

Keep a problem-solving journal in which you *(a)* record interesting problems (including some not yet solved), *(b)* describe what strategies you used or thought about, *(c)* explain how you verified solutions (e.g., checked the conditions, reworked the problem in another way, checked the reasonableness of solutions), *(d)* identify similar or related problems, and *(e)* record problems posed by other students.

A problem-solving journal requires students to reflect on what they do when they solve problems. It can provide information on all aspects of problem solving. Teachers will detect progress if students increase the number of strategies they report using, verify their solutions in different ways, and find relationships among problems. The development of a language to talk and write about problems also should be evident in students' increasing use of terms describing strategies and related problems.

Grades 9—12

1. *Formulate problems; solve problems.*

a. **You have 10 items to purchase at a grocery store. Six people are waiting in the express lane (10 items or fewer). Lane 1 has one person waiting, and lane 3 has two people waiting. The other lanes are closed. What check-out line should you join?**

b. **You are considering purchasing one of two cars, both four years old. One car costs $3000 and gets 20 miles a gallon. The other costs $4500 and gets 35 miles a gallon. Which car is the best buy if you plan to keep it two years?**

What additional information do you need to answer these questions?

One aspect of formulating problems is identifying whether additional information is needed. Neither of the problems above provides all the information needed to make a decision. Students need to identify the missing information and the likely estimates for the missing quantities. In question *a*, the number of items each person has and the speed of the checkers are considerations. In problem *b* the number of miles traveled each year, the price of gasoline, and cash available are considerations. If money has to be borrowed to purchase the more expensive car, it can make a difference. These problems are appropriate for individual or small-group work embedded in instruction. Notes can be kept on the variety of questions generated and what additional information is assumed. In class, the willingness of students to engage themselves in finding the necessary information can be observed. Calculators are important for question *b*.

2. *Solve problems; generalize solutions.*

Prove each of these statements:

a. **The sum of two consecutive whole numbers is not divisible by 2.**

b. **The sum of three consecutive whole numbers is divisible by 3.**

State what you consider to be the general case for the statements. Prove or give a counterexample.

♦ ♦ ♦ ♦ ♦ ♦ ♦ ♦

This task can be included on a test or as a take-home problem. Evaluating this task includes assessing students' ability to prove the statements; the nature of the statement developed; and their ability to prove or disprove the general statement. A possible statement of the general case is this:

Is it true that the sum of an even number of positive consecutive whole numbers is not divisible by the number but that the sum of an odd number of positive consecutive whole numbers is divisible by the number?

This problem can be scored by giving points for the different parts of the problem solution—one point for accurately representing the first statement [$(n + (n + 1))/2$ = an integer?], one point for giving a convincing argument for two consecutive whole numbers, one point for accurately representing the second statement, one point for giving a convincing argument for that statement, two points for stating the general case, two points for providing an adequate proof, and two points for explaining what strategies were used.

STANDARD 6: COMMUNICATION

The assessment of students' ability to communicate mathematics should provide evidence that they can—

- ***express mathematical ideas by speaking, writing, demonstrating, and depicting them visually;***
- ***understand, interpret, and evaluate mathematical ideas that are presented in written, oral, or visual forms;***
- ***use mathematical vocabulary, notation, and structure to represent ideas, describe relationships, and model situations.***

Focus

The Curriculum Standards present a dynamic view of the classroom environment. They demand a context in which students are actively engaged in developing mathematical knowledge by exploring, discussing, describing, and demonstrating. Integral to this social process is communication. Ideas are discussed, discoveries shared, conjectures confirmed, and knowledge acquired through talking, writing, speaking, listening, and reading. The very act of communicating clarifies thinking and forces students to engage in doing mathematics. As such, communication is essential to learning and knowing mathematics. But communicating mathematically presents unique difficulties for students. Mathematics is heavily based on the use of symbols and attaches specific, and sometimes different, meanings to common words. The result can be confusion and difficulty in expressing mathematical ideas. Traditional forms of testing cannot always identify such confusion and difficulty, and thus they ignore the social context of mathematics.

An assessment of students' ability to communicate mathematically should be directed at both the meanings they attach to the concepts and procedures of mathematics and their fluency in talking about, understanding, and evaluating ideas expressed in mathematics. The evaluation should include different forms of communication and should emphasize communication both among people and with various forms of technology. Assessment also must be sensitive to students' language development. As in any language, communication in mathematics means that one is able to use its vocabulary, notation, and structure to express and understand ideas and relationships. In this sense, communicating mathematics is integral to knowing and doing mathematics.

Discussion

Since communication is a social activity that takes place within a context, it should be assessed in a variety of situations. In assessment, as in teaching, teachers should be aware of how students express mathematical ideas and how they interpret the mathematical expressions of others. In assessing students' ability to communicate, teachers should pay attention to the clarity, precision, and appropriateness of the language used. In addition, students' ability to understand the written and oral communication of others is an important component of instruction and assessment.

◆ ◆ ◆ ◆ ◆ ◆ ◆ ◆

Grades K–4

In the early grades, when students are introduced to mathematical vocabulary and notation, the assessment of their understanding of language is important as they form initial conceptions of the subject. In many situations, unfamiliar terms and phrases are encountered, and familiar terms are used in unfamiliar ways. For example, although children have often heard the question *How many?* the phrase *How many more?* suggests a quantitative relationship and might confuse them. At the same time, they are learning the meaning and correct use of symbols, some of which denote operations whereas others denote relationships. An assessment sensitive to this confusion will identify the possible causes of errors while it varies the format of the problems. A child who gives 12 as the answer to $5 + \square = 7$, for example, should be questioned further to determine if the equal sign is used as a cue for addition rather than as the symbol signifying that the quantities on each side are equivalent. In this example, the difficulty might lie in an understanding of the language rather than of the operation.

At these ages, the assessment of communication should occur informally in the context of instruction. Although children should be encouraged to verbalize their thoughts, asking for general explanations or explicit meanings of a concept or procedure can be threatening. Children at these ages are more comfortable describing these processes with specific examples, such as making a drawing of a number problem or demonstrating a procedure through multibase blocks. At the same time, they should be encouraged to verbalize their thinking so that their development of language, along with the concepts, can be monitored. Their willingness to participate in class discussion also should be noted. The degree to which children are comfortable in expressing their mathematical thinking and their flexibility in using various forms of communication are primary aspects of communication at this level.

It is also important that children be evaluated on how well they listen. In an environment in which children are encouraged to question and challenge, the teacher will find ample indications of how well children have interpreted and evaluated what has been said or presented. The posing of such specific questions as "How does Jim's solution differ from Susan's?" or such general ones as "Can you say that in your own words?" will reveal the extent to which a student has understood the message.

The following is an example of a task for assessing students' speaking and listening skills:

Two students each have a geoboard. One geoboard has a design, and the other is blank. Each student is seated so that he or she cannot see the other's geoboard. The student with the design gives the other student directions on how to reproduce the design. See figure 6.1.

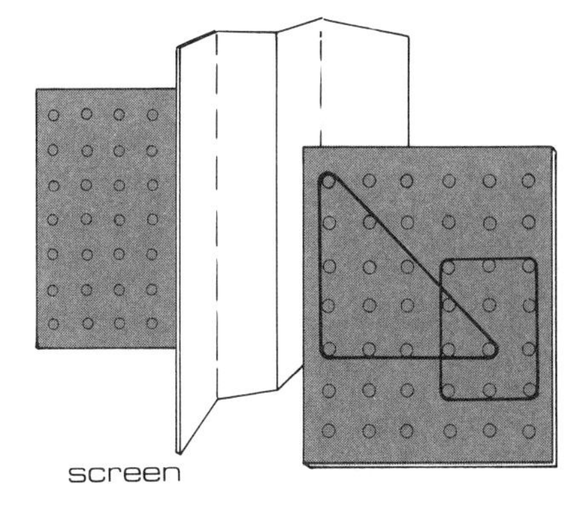

***Fig. 6.1.* Arrangement of geoboards for assessing speaking and listening skills**

A variation of this task is to let the second student ask questions. In an evaluation of students' communication skills, the accuracy of the reproduced design is one indication of their interchange. Other observations can focus on whether the student talked about the figure by breaking it into parts (vertical line, horizontal line, three units) or by identifying the whole figure (right triangle, square) and if the first student repeated an instruction in more than one way. The opportunity to observe the students' communication skills continues as the students compare the two designs and talk about their similarities and differences. Their vocabulary and if they described the design in more than one way should be noted.

Grades 5–8

In the middle grades, as students become more aware that the meanings of many mathematical terms and notations depend on their context, assessing the clarity of communication is vital. Because students tend to overgeneralize earlier meanings, special attention should be paid to their understanding of terms and operations. For example, previous experiences with whole numbers lead students to view the effects of multiplication as an increase in quantity. This conception must now be modified to include the effects of multiplication on fractions and negative numbers. Similar extensions of meaning occur with symbols— for example, when a "−" sign that was originally encountered in subtraction is now used to denote quantities with magnitude and direction (e.g., $3 - (-4)$).

As in the earlier grades, the assessment of students' ability to understand mathematical terms and concepts is best achieved through a natural extension of instructional activities. Such questions as "Why?" "What if?" and "How would you convince someone?" should be asked routinely to help students explain or justify their answers or conjectures. Here, the criterion for acceptability should be based more on the clarity of presentation than on the precision of mathematical vocabulary. Again, assessment should usually be oral and informal. When students accept the communication of ideas as a normal part of their lessons, they will be more willing to express them.

It also is important to assess how well students can interpret both the verbal and the visual communication (e.g., graphs) of others. Students should be able to interpret and extrapolate from graphs as well as read them. In turn, they should be able to discuss and defend their interpretations. For example, if one group of students used the box plots in figure 6.2 when comparing findings from a survey of the number of hours of TV watched each day by students in their school with the results of a national survey (see 5–8 Standard 2), another group can be asked to interpret the data. The ability of the second group to describe the central points and the distribution of the data and justify their conclusions can be assessed.

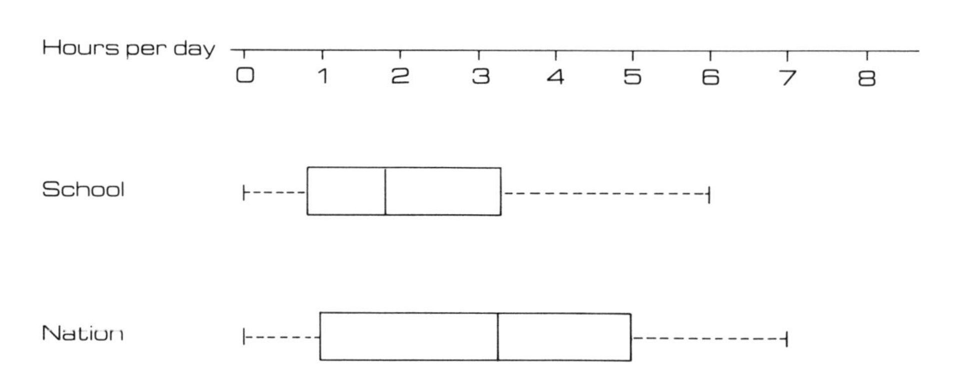

Fig. 6.2. Box plots of hours of TV watched by students for school and nation

Grades 9–12

At the secondary school level, students encounter more abstract ideas and experience in greater depth the formal language of mathematics. Assessment should be directed toward students' understanding of the mathematical language, its terms and syntax, as well as toward their appreciation of the role of rigor and precision in communicating mathematical ideas. Although informal observation continues to be important at the high school level, the increase in formal mathematical presentation requires new criteria for assessment. At this level, students' written work

♦ ♦ ♦ ♦ ♦ ♦ ♦ ♦

should be judged for its precision, clarity, and the appropriateness of the presentation. Students should be able to form multiple representations of ideas and relationships and recognize their relative appropriateness. However, expectations for the use of symbols should relate to the maturity of the students and the context of the task. At times, a symbolic expression might be required; in other contexts, a mixture of symbols and natural language might be adequate; in others, symbolism might be completely unwarranted.

It is important to note that students' ability to communicate mathematically can be assessed by having students write about mathematics. The written responses should be judged for accuracy, clarity, precision, and the proper use of mathematical terms and symbols. The following is a sample task:

Imagine you are talking to a student in your class on the telephone and want the student to draw some figures. The other student cannot see the figures. Write a set of directions so that the other student can draw the figure and graph exactly as shown in figure 6.3.

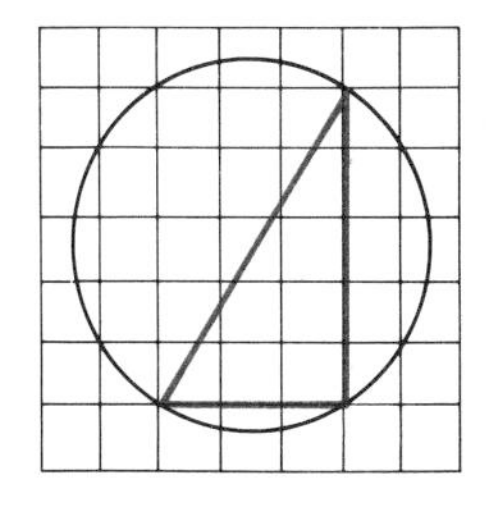

(a)

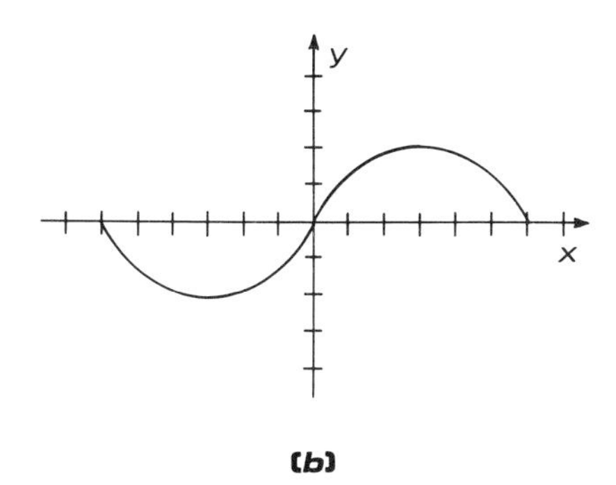

(b)

***Fig. 6.3.* Figures for writing a set of directions to reproduce drawings**

Assessment should include more than judgments of written work. Although students' ability to understand mathematics texts or articles can be assessed through written summaries, discussion can be a more useful context for judging students' ability to function as active, critical participants in the reading or listening processes occurring during class or small-group discussions.

The assessment of students' ability to communicate through technology is also important. The increasing use of technology as a tool demands that students be able to use computers and software, such as spreadsheets and data-base programs, to structure and present information. The criterion for performance is whether students can communicate using technology. One way of assessing this ability is by determining if students can use a spreadsheet to simulate a situation and provide evidence for a conclusion. For example:

Using an electronic spreadsheet, demonstrate that if 1 gallon of deicer fluid is added with each fill-up of a fuel tank, a limit of 2 gallons of deicer in the tank at the fill-up will eventually be reached. Assume that the driver habitually fills the tank when it is half full.

A student who can adequately communicate through a computer can use one column to designate the number of fill-ups and a second column to report the amount of deicer. To generate a value in the second column, one uses the sum of a geometric series with the first term 1 and a ratio of 0.5. An example of a successful response to this task is given in figure 6.4 (Day and Scott 1987) on the next page.

```
   I  A  II  B  II  C  II  D  II  E  I
 1
 2 This model simulates the addition of
 3 gasline deicer to a car's fueltank,
 4 always adding the same amount of deicer
 5 at the same point of "emptiness" of the
 6 fuel tank.  Enter appropriate values in
 7 cells A10, A11, and A12.
 8 ****************************************
 9                                     ****
10  14.5    TANK CAPACITY (GALLONS)    ****
11  1       AMOUNT DEICER ADDED (GAL)  ****
12  .5      PART TANK FULL             ****
13 ****************************************
14    FILL  AMOUNT OF  RATIO
15  NUMBER  DEICER     DE/GAS
16 ===========================
17  1        1         .0740741
18  2        1.5       .1153846
19  3        1.75      .1372549
20  4        1.875     .1485149
21  5        1.9375    .1542289
22  6        1.96875   .1571072
23  7        1.984375  .1585518
24  8        1.992188  .1592755
25  9        1.996094  .1596376
26  10       1.998047  .1598188
27  11       1.999023  .1599094
28  12       1.999512  .1599547
29  13       1.999756  .1599773
30  14       1.999878  .1599887
31  15       1.999939  .1599943
32  16       1.999969  .1599972
33  17       1.999985  .1599986
34  18       1.999992  .1599993
35  19       1.999996  .1599996
36  20       1.999998  .1599998
37  21       1.999999  .1599999
38  22       2.000000  .1600000
39  23       2.000000  .1600000
40  24       2.000000  .1600000
41  25       2.000000  .1600000
```

Fig. 6.4. **Spreadsheet simulation of repeated fill-ups of an automobile's fuel tank**

STANDARD 7: REASONING

The assessment of students' ability to reason mathematically should provide evidence that they can—

- ***use inductive reasoning to recognize patterns and form conjectures;***
- ***use reasoning to develop plausible arguments for mathematical statements;***
- ***use proportional and spatial reasoning to solve problems;***
- ***use deductive reasoning to verify conclusions, judge the validity of arguments, and construct valid arguments;***
- ***analyze situations to determine common properties and structures;***
- ***appreciate the axiomatic nature of mathematics.***

Focus

The types of reasoning identified in this standard are fundamental to doing mathematics but cannot always be observed in students' verbal answers or written work. It is natural for students to form conjectures on the basis of the examples they have seen or worked and to develop arguments that are based on what they know to be true. Students can also have intuitive notions about proportional reasoning and spatial relationships. All students should have explicit opportunities to engage in such intuitive, informal reasoning and, hence, any assessment of students' reasoning abilities should obtain evidence of these processes.

Accordingly, assessment techniques should specifically assess students' use of different types of reasoning. Although some aspects of reasoning might be more appropriate than others at a given grade level, all aspects can be used at any grade level. However, in the early grades, some aspects may be used only in an intuitive sense.

Discussion

The following examples illustrate tasks or activities for assessing students' abilities to reason. They can be used in the context of instruction, such as in class discussions, or as formal assessment tasks. Young students should be asked to discuss or explain their answers orally.

Grades K–4

1. *Deductive reasoning using known facts*

If $35 - 20 = 15$, what is $35 - 19$? Why?

Again, students should explain their reasoning processes. Of interest is the student's ability to use the first numerical equation to develop the second. A response that relies only on the recall of facts and concentrates solely on the second equation in isolation from the first does not demonstrate deductive reasoning.

♦ ♦ ♦ ♦ ♦ ♦ ♦ ♦

2. *Analyzing a situation to determine common properties and structures*

Ask students to work in small groups with cutouts of squares and rectangles. Ask each group to consider questions such as these:

a. **What properties do squares and rectangles have in common?**
b. **What properties do they not have in common?**

In a strong response, students would compare the properties of both figures simultaneously, recognizing that both figures have four right angles. Less sophisticated students might first list the properties of a square and then of a rectangle but would fail to make comparisons between the two figures. Clearly the task requires conceptual understanding—in particular, the ability to compare and contrast concepts.

3. *Spatial reasoning*

Blindfold children and have them handle a cube and a square pyramid. Have each child consider questions such as these:

a. **What figure are you holding?** ***b.*** **How many "corners" (vertices) does it have?**

This activity can be extended by including other solids or by posing other questions about the cube and the square pyramid, such as, "How many lines [edges] do the solids have?" Students who can provide more detailed descriptions of the solids are demonstrating better spatial reasoning than those who seem to rely on a rather "mechanistic" counting of vertices, edges, or faces.

Grades 5–8

1. *Inductive reasoning*

Ask students to consider the following situation:

Five students have test scores of 62, 75, 80, 86, and 92. Find the average score. How much is the average score increased if each student's score is increased by—

a. **1 point?**
b. **5 points?**
c. **8 points?**
d. ***x*** **points?**

Write a statement about how much the average score is increased if each individual score is increased ***x*** **points. Develop an argument to convince another student that the statement is true.**

Some students might need more than three instances to form a conjecture. This exercise focuses on a student's ability to generalize from specific cases. Students who can find the answer for each specific problem but who are unable to state the general case are less able to use inductive reasoning than those who can state the general case after three or four instances. Students can work in small groups. A computer can be used to investigate situations other than the three posed or to consider the scores of an entire class.

◆ ◆ ◆ ◆ ◆ ◆ ◆ ◆

2. *Deductive reasoning and developing a plausible argument*

Ask small groups of students to construct models like those in figure 7.1 and develop the formula for the area of a circle or explain why the formula $A = \pi r^2$ is plausible.

Students who can use the relationship between the shape of the "parallelogram" and its area and the circumference of the circle to develop the formula for the area of the circle are demonstrating plausible and deductive reasoning. The argument is plausible if it makes common sense and is mathematically correct.

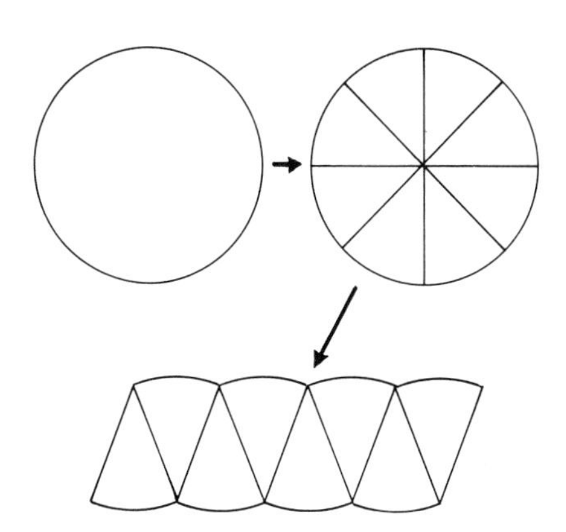

Fig. 7.1. Model relating the area of a circle to the area of a parallelogram

3. *Proportional reasoning*

Pose this question: How many students in the school are left-handed? Have students develop a procedure in which they examine a sample of students for left-handedness and use proportional reasoning to determine the number of students in the entire school who are left-handed.

Students would need to collect data and set up and solve a proportion to answer this question. Other topics can be investigated in a similar way. Small groups of students can gather information about the community at large—for example, how many people are left-handed—as part of a long-term project.

Grades 9—12

1. *Spatial reasoning*

What formula(s) can be used to find the area of each of the cross-sections of a cube containing the points indicated in figure 7.2?

The evaluation process should consider several steps: First, the students need to visualize the plane containing the indicated points. Second, they must be able to describe the shape of the cross-section. Finally, they need to identify the desired area formulas. Students can work in small groups or individually.

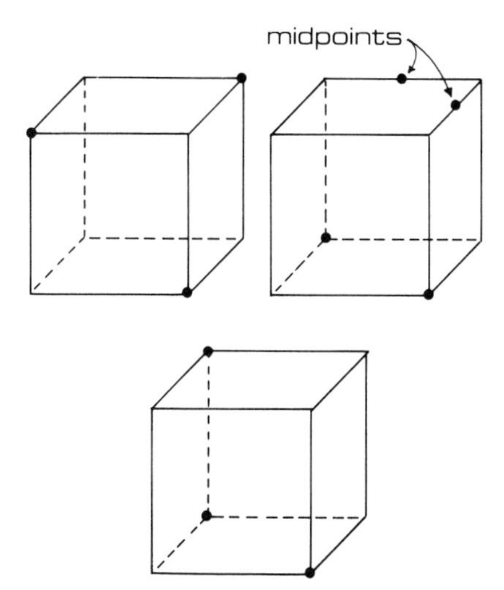

Fig. 7.2. Points indicating cross-sections of cubes for which formulas are to be found

2. *Deductive reasoning*

Roger doesn't believe that adding the same number of points to each student's test score will increase the average score by that same amount. Write a valid argument to convince Roger that this is true.

This argument should be deductive: A specific case or several cases are not sufficient. Some students might select a specific increase (say, five points) and argue that case. Because of its specificity, this argument is not as strong as selecting a general increase (n points) and showing that the average increases by n points.

3. *Appreciating the axiomatic nature of mathematics*

Students at all levels must develop an intuitive sense that mathematics is based on established rules and is not a "bag of tricks" familiar only to those who teach or develop mathematics. It is particularly important that advanced students understand that there is an element of arbitrariness in how the rules are selected but that the encompassing system is consistent. The following problems can be given to geometry students, since one of the desired outcomes of teaching geometry is that students develop a sense of what constitutes an axiomatic system.

***a.* Write an essay on the following topic: In what way did the mathematicians who developed non-Euclidean geometry contribute to the notion that postulates can be arbitrarily selected in mathematics?**

The essay should focus on how the parallel postulate (Euclid's fifth postulate) cannot be proved on the basis of preceding postulates and theorems and that there are different options in defining a parallel postulate. The axiomatic nature of mathematics should be highlighted. Students can work in small groups to develop an essay, which can then be presented in class.

***b.* Suppose a mathematical system assumes the following statement as a postulate:**

If two lines are parallel and cut by a transversal, the alternate interior angles are congruent.

Suppose it proves the following statement as a theorem:

If two lines are parallel and cut by a transversal, the corresponding angles are congruent.

Is it possible to assume that the second statement is a postulate and prove the first statement as a theorem? Why?

The student's response should emphasize that the selection of the postulate is arbitrary but that once selected, the second statement can be proved on the basis of the accepted postulate.

STANDARD 8: MATHEMATICAL CONCEPTS

The assessment of students' knowledge and understanding of mathematical concepts should provide evidence that they can—

- ***label, verbalize, and define concepts;***
- ***identify and generate examples and nonexamples;***
- ***use models, diagrams, and symbols to represent concepts;***
- ***translate from one mode of representation to another;***
- ***recognize the various meanings and interpretations of concepts;***
- ***identify properties of a given concept and recognize conditions that determine a particular concept;***
- ***compare and contrast concepts.***

In addition, assessment should provide evidence of the extent to which students have integrated their knowledge of various concepts.

Focus

Concepts are the substance of mathematical knowledge. Students can make sense of mathematics only if they understand its concepts and their meanings or interpretations. For example, if students are to make sense of the procedure for subtracting whole numbers with regrouping, they must understand the concept of place value. Likewise, if students are to recognize that a given situation calls for subtraction, they must understand the concept of subtraction and recognize that the action depicted in the situation corresponds to one of its meanings (e.g., take away, comparison, partition). Because conceptual understanding is fundamental to doing mathematics meaningfully, an assessment of students' knowledge must examine their grasp of mathematical concepts.

An understanding of mathematical concepts involves more than mere recall of definitions and recognition of common examples; it encompasses the broad range of abilities identified in this standard. Assessment, too, must address these aspects of conceptual understanding. Assessment tasks should focus on students' abilities to discriminate between the relevant and the irrelevant attributes of a concept in selecting examples and nonexamples, to represent concepts in various ways, and to recognize their various meanings. Tasks that ask students to apply information about a given concept in novel situations provide strong evidence of their knowledge and understanding of that concept. Problems designed to elicit information about students' misconceptions can provide information useful in planning or modifying instruction.

Discussion

The assessment of students' understanding of concepts should be sensitive to the developmental nature of concept acquisition. Students' grasp of mathematical concepts develops over time. Many concepts introduced in the early grades are later extended and studied in greater depth. A fraction, for example, is introduced as a part of a whole in the primary grades, as a measure in the intermediate grades, and as a ratio in junior high school; finally, algebraic fractions are taught in secondary school. This progression is accompanied by the development of the language and notation of fractions and extended further through the exploration of the relationships among fractions and other concepts, such as decimals, and through the applications of fractions in various contexts, such as in proportional reasoning. Tasks and situations used to measure the understanding of concepts should change over the grades to determine whether students' notions of the concept are maturing.

This standard suggests the kinds of tasks needed to assess the various aspects of students' conceptual understanding and knowledge. As the following examples illustrate, it is unnecessary to assign separate tasks to assess each aspect of understanding; it is feasible—in fact, advisable—to design a single task that covers several aspects. Nor is it necessary to assess all aspects of all concepts or all students at all times. The aspects of conceptual understanding to be assessed should be selected according to the mathematical content and the level of the students. Furthermore, assessment tasks should be consistent with the methods of instruction. For example, children in the early grades, whose understanding of fractions is closely tied to the use of physical materials, should be encouraged to use such materials to demonstrate their conceptual knowledge. The examples are arranged by grade level, but many are appropriate for students in grades other than the ones specified. Similar problems can be created for different levels of involvement (whole class, small groups, or individual) and mode of response (written, oral, performance, or computer).

Grades K—4

Fig. 8.1

1. *Identify examples and nonexamples of concepts. (See fig. 8.1.)*

Which figures show that exactly 1/2 of the region is shaded?

Here, students are to identify examples and nonexamples of one-half of a region. The task is constructed to determine whether they can discriminate between the relevant and irrelevant attributes (i.e., the area of the shaded region must be equal to that of the unshaded region, but the two parts need not be contiguous or congruent).

2. *Use models to represent concepts.*

A yellow hexagonal pattern block represents one whole. Use blocks to represent 1/2, 1/3, 1/6, 2/3, and so on.

Students should be asked to explain their responses and encouraged to show a given fractional part in more than one way, for example, showing 1/2 using one trapezoidal block or three triangular blocks. Various blocks or combinations of blocks can represent one whole. In this context, students who consistently show fractional parts of a given whole demonstrate a good understanding of fractions as parts of regions. Students can complete this activity individually, in small or large groups, or in a multiple-choice format in which they select the block that represents a particular fraction.

♦ ♦ ♦ ♦ ♦ ♦ ♦ ♦

3. *Identify properties of given concepts; compare and contrast concepts.*

Give students a sheet containing drawings of various quadrilaterals. (Include drawings of shapes in various orientations. Drawings should be numbered for easy reference.) Ask students to cut out the shapes and hold up those that fit the following descriptions:

Hold up shapes with two pairs of parallel sides. What are they called? (Parallelograms; responses should include squares and rectangles.)

Hold up shapes with four congruent sides. What are they called? (Rhombuses; responses should include squares in various orientations.)

This task requires students to identify shapes that have specified characteristics. From the shapes displayed, it can be determined quickly which children recognize the properties of various quadrilaterals and their interrelationships. For example, children who display only parallelograms that are not rectangles in response to the first question do not have a full understanding of parallelograms, rectangles, and squares. It is important that children be asked to explain their choices as a further means of assessment.

4. *Integrate knowledge of concepts.*

Direct students to work in groups of four or five. Give each group three paper "pizzas." Ask them to determine how much pizza each group member will receive if they share their pizzas equally.

This task can yield information about students' understanding of fractions as quotients and about their ability to use what they know about fractions in a novel situation. Students who can describe the amount of pizza each person receives as a fractional part of a whole pizza demonstrate a good understanding of fractions as parts of a region.

Grades 5–8

1. *Recognize various interpretations of concepts; use diagrams to represent concepts; translate between modes of representation.*

Complete each diagram in figure 8.2 so that it shows 2/5.

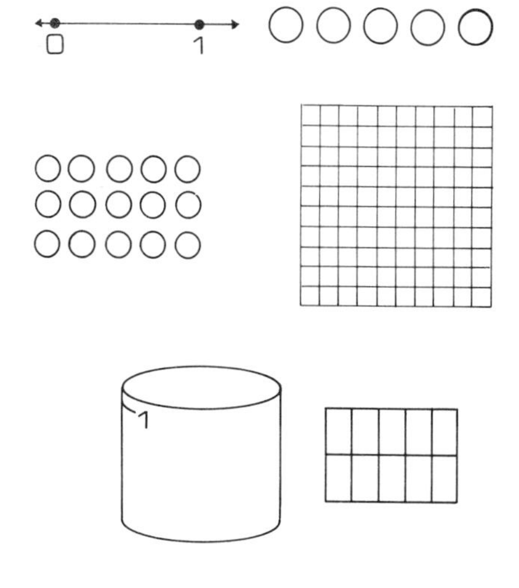

Fig. 8.2

This task assesses students' recognition of the various meanings and interpretations of a given fraction. It also requires them to translate between symbolic and pictorial modes of representation and to use diagrams to represent concepts. Students who can represent a given fraction in all these contexts demonstrate a broad understanding of the meaning of fractions. For seventh- and eighth-grade students, this task can be extended to representations of fractions as decimals, percents, and ratios.

2. *Identify examples and nonexamples of concepts; compare and contrast concepts.*

In figure 8.3, put a Q on each shape that is a quadrilateral; a P on each parallelogram; an R on each rectangle; an RH on each rhombus, and an S on each square. You can put more than one letter on a single shape.

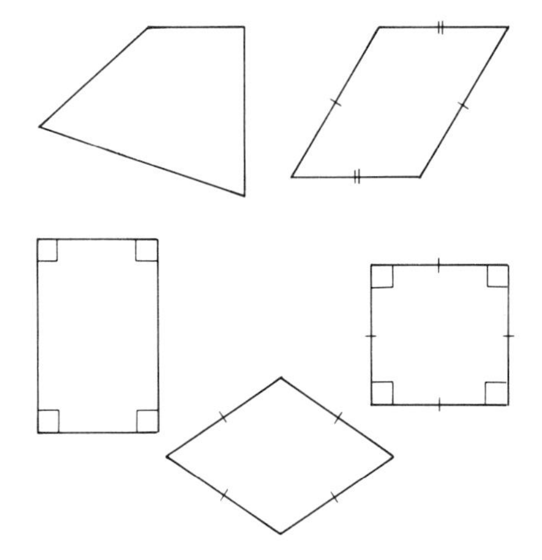

Fig. 8.3

Students who correctly identify each figure are demonstrating that they know a single shape can be classified in several ways. This evidence shows that they understand class-inclusion relations among types of quadrilaterals (i.e., that all squares are rectangles and rhombuses and that all rectangles are parallelograms). This activity can be extended by asking students to justify their responses, orally or in writing, by completing the statement, "A quadrilateral is a shape that . . . ." Similar activities can determine whether students can recognize relevant and irrelevant attributes of types of shapes (e.g., that orientation is irrelevant).

3. *Integrate concepts.*

On a journey from Pittsburgh to New York, Pat fell asleep after half the trip. When she awoke, she still had to travel half the distance that she traveled while sleeping. For what part of the entire trip had she been asleep?

Assuming that the shaded part in each diagram in figure 8.4 shows when Pat was asleep, which diagram best depicts the answer to the problem?

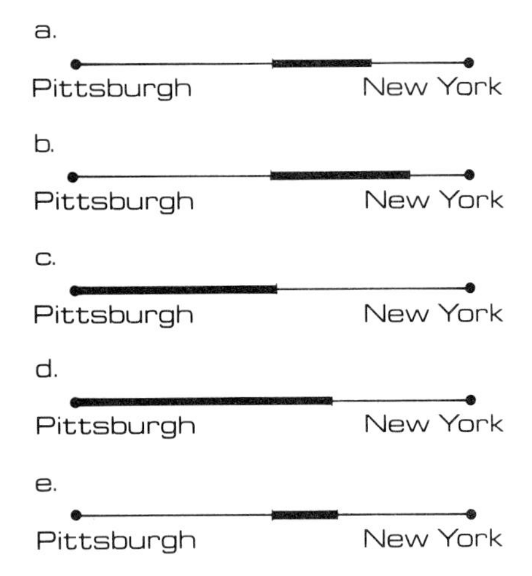

Fig. 8.4

This task assesses whether students can use their knowledge of fractions to interpret the problem and identify the correct representation of its solution. It requires that they consider a fraction in relation to different units. They must first think of the whole trip as a unit, then consider a different unit (the portion of the total trip during which Pat was asleep).

Grades 9–12

1. *Identify examples and nonexamples of concepts.*

Which of the following represent rational numbers?

$2/3$ $\quad$ $\sqrt{4/5}$ $\quad$ 0 $\quad$ $\sqrt{5}$ $\quad$ 1.3434 $\quad$ -5.6

$1.121121112...$ $\quad$ $\sqrt{-16}$ $\quad$ $7/(9-3^2)$ $\quad$ $-6/-2$ $\quad$ 25%

Again, the task is designed to determine whether students can discriminate between relevant and irrelevant attributes of rational numbers. Students who identify $2/3$, $\sqrt{4/5}$, $7/(9-3^2)$, and $-6/-2$ as rationals might be confusing fraction with rational number; those who classify $1.121121112...$ as rational might not be distinguishing repeating decimals from nonrepeating decimals containing patterns. This task can be presented in many formats, including a written exercise in which students are asked to justify their selections.

2. *Recognize conditions that determine a concept.*

Who am I?

I am an equiangular quadrilateral. What special kind of quadrilateral am I?

Here, students must distinguish between properties that some quadrilaterals possess and properties that are sufficient to define them. For example, to answer this item correctly, students must recognize that al-

◆ ◆ ◆ ◆ ◆ ◆ ◆ ◆

though the angles of squares are equal, all rectangles have this property, so the correct answer must be rectangles, not squares.

3. *Translate from one mode of representation to another.*

What is the equation of the line shown in figure 8.5?

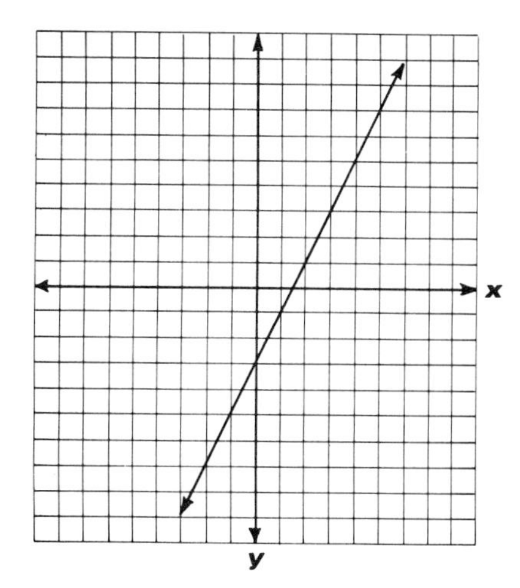

Fig. 8.5

This task requires students to translate a graphic representation of a line into a symbolic representation. To do so efficiently, students must be able to determine the slope and intercept of the line from its graph, then represent them symbolically to create the desired equation, $y = 2x - 3$. Students who are able to do such re-representations with ease demonstrate a solid understanding of the concept of line in analytic geometry and of the related concepts of slope and intercept.

4. *Integrate concepts.*

Connect the midpoints of the four sides of an isosceles trapezoid. What kind of figure do you get? Justify your answer.

This task yields information about the extent to which students have integrated their knowledge of geometric concepts. To solve this problem, students must be able to draw an isosceles trapezoid, find the midpoints of the sides, and then recognize the figure that results from connecting the midpoints. Students must then be able to apply additional knowledge about the conditions that determine whether a figure is a rhombus to justify their answers. Students might be able to identify the desired figure yet be unable to justify their responses.

STANDARD 9: MATHEMATICAL PROCEDURES

The assessment of students' knowledge of procedures should provide evidence that they can—

- ***recognize when a procedure is appropriate;***
- ***give reasons for the steps in a procedure;***
- ***reliably and efficiently execute procedures;***
- ***verify the results of procedures empirically (e.g., using models) or analytically;***
- ***recognize correct and incorrect procedures;***
- ***generate new procedures and extend or modify familiar ones;***
- ***appreciate the nature and role of procedures in mathematics.***

Focus

In the context of school mathematics, procedures generally mean computational methods. But not all procedures in the mathematics curriculum are computational. Geometric constructions, such as bisecting an angle and constructing the perpendicular to a line at one of its points, are procedural but not computational. The various aspects of procedural knowledge identified in this standard apply equally well to noncomputational procedures.

Although it is important that students know how to execute mathematical procedures reliably and efficiently, a knowledge of procedures involves much more than simple execution. Students must know when to apply them, why they work, and how to verify that they give correct answers; they also must understand concepts underlying a procedure and the logic that justifies it. Procedural knowledge also involves the ability to differentiate those procedures that work from those that do not and the ability to modify them or create new ones. Students must be encouraged to appreciate the nature and role of procedures in mathematics; that is, they should appreciate that procedures are created or generated as tools to meet specific needs in an efficient manner and thus can be extended or modified to fit new situations. The assessment of students' procedural knowledge, therefore, should not be limited to an evaluation of their facility in performing procedures; it should emphasize all the aspects of procedural knowledge addressed in this standard.

Discussion

It should be evident that procedural knowledge is intertwined with conceptual knowledge. For example, one cannot extend or modify a procedure for finding the least common multiple of two numbers unless the concept of common multiple is itself understood. Thus, the examples that follow concern aspects of both conceptual and procedural knowledge. They focus on both computational and noncomputational procedures across grade levels and describe tasks for assessing the various aspects of procedural knowledge.

♦ ♦ ♦ ♦ ♦ ♦ ♦ ♦

Grades K—4

1. *Recognize when to use a procedure.*

Divide the class into small groups. Direct each group to create story situations containing two-digit numbers with some involving multiplication. The groups then can exchange problems and identify those that require the multiplication of two-digit numbers.

The problems may differ with respect to the cleverness of the story or the context of the multiplication. Some possibilities include additive situations that suggest multiplication—Mike ate 11 strawberries for 23 days. How many did he eat?—or multiplicative situations—Wanda had 12 skirts and 15 blouses. How many possible outfits did she have? Solving more sophisticated problems can require more than one step: Curtis ran 6 miles every day and Valerie ran 5 miles every day. How many total miles did they run in two weeks? Assessment can focus on such considerations as whether the problems call for multiplication as requested, the richness of the situation, the meaning of multiplication in the problem, the ability of students to discriminate between problems that call for multiplication and those that do not, and whether the problem makes sense regardless of the procedure involved.

2. *Verify the results of a procedure.*

Solve 62 − 35. Use multibase blocks or other materials that can represent two-digit numbers to show that your answer is correct.

Each student should verify the subtraction process individually. Verification can involve regrouping the blocks or counting them. The purpose of the activity is to have students use a procedure to find the answer and then show empirically that their procedure works. Assessment can focus on the students' ability to interpret regrouping or to demonstrate how they used a counting technique; in any event, students should be encouraged to offer a more complete explanation than a simple description of the mechanical process of subtraction.

3. *Generate a new procedure.*

Anita is trying to find a way to solve two-digit subtraction problems like 75 − 26 without regrouping. How can she change the problem so that the answer will be the same and she will not have to regroup 75 into 60 + 15?

One method is to change the problem to 79 − 30 (add 4 to both numbers) or to 69 − 20 (subtract 6 from each number). Whatever method students use, it is important that they explain how the new procedure changes the numbers in the original task without changing the difference. Some students might try to process the problem by using different counting mechanisms. Assessment should focus on the accuracy of the selected procedure and the student's ability to explain why it works.

Grades 5—8

1. *Recognize when to use a procedure.*

The lockers in Pythagoras Middle School are numbered from 1 to 500. Starting with locker 1, we find that every sixth locker has a blue de-

♦ ♦ ♦ ♦ ♦ ♦ ♦ ♦

cal, every ninth locker has a yellow decal, and every tenth locker has a green decal. What is the number of the first locker to have all three decals?

Solving this problem requires finding the least common multiple (LCM) of 6, 9, and 10. Hence, students will need to use a procedure or rely on their conceptual understanding to produce a set of multiples and then select the least of those. In either event, the solution of the problem has two parts: (1) recognizing that an LCM must be found and (2) correctly finding the LCM of 6, 9, and 10. An assessment should consider both parts of the problem.

2. *Reliably and efficiently execute procedures.*

Find the least common multiple of the following numbers:

***a.* 12, 18 *b.* 7, 21 *c.* 8, 9**
***d.* 1, 6 *e.* 6, 9, 10 *f.* 5, 6, 20**

Note the variety of numbers given. In set *b,* one number is the multiple of the other. In set *c,* the numbers are relatively prime. In set *d,* one of the numbers is 1. In sets *e* and *f,* the three numbers might require some modification of the procedure that students usually use. The assessment should focus on whether students can arrive at the correct answer with reasonable proficiency.

3. *Recognize correct and incorrect procedures.*

Hershel was given the problem 2/5 $<$? $<$ 4/7. He said that 3/6 would be between 2/5 and 4/7. The teacher asked Hershel to explain how he got his answer and why he thinks his method works. Hershel said that he chose a numerator of 3 because 2 $<$ 3 $<$ 4 and a denominator of 6 because 5 $<$ 6 $<$ 7. Hershel claimed his method always works and gave the following examples:

***a.* The fraction 2/4 is between 1/3 and 3/5.**

***b.* The fraction 4/9 is between 2/5 and 6/11.**

Are Hershel's examples correct? Does his procedure always work? Explain your reasoning.

Small groups of students can generate other examples to test whether Hershel's procedure works. A key step in the problem is considering various pairs of fractions and a wide array of possibilities for the "new" numerator and denominator. Assessment should focus on whether students can generate examples that fit Hershel's model and on their cleverness in selecting those examples. If they continually choose examples in which the numerators and denominators differ by 2, they might incorrectly conclude that the procedure works. The identification of a fraction that doesn't work, such as 3/8 in example *b,* is a valid response. A more insightful response is to consider what the procedure means when it is applied to fractions like 1/3 and 2/4.

Grades 9–12

1. *Give reasons for the steps in a procedure.*

Justify each of the following steps in multiplying (x + 4) by (x + 2):

♦ ♦ ♦ ♦ ♦ ♦ ♦ ♦

$$
\begin{aligned}
(x + 4)(x + 2) &= x(x + 2) + 4(x + 2) \\
&= x^2 + 2x + 4x + 8 \\
&= x^2 + (2 + 4)x + 8 \\
&= x^2 + 6x + 8
\end{aligned}
$$

Reasons can be explained orally or in writing. Assessment should focus on how articulate the students are in providing mathematical reasons (axioms, definitions, theorems) for each of the steps.

2. *Verify the results of a procedure.*

a. **Find the inverse of $A = \begin{bmatrix} 3 & -1 \\ 1 & 4 \end{bmatrix}$.**

How can you verify that your new matrix is the inverse of *A*?

b. **Draw a line segment and trisect it using a compass and straightedge. Use paper folding to verify that the segment has been trisected.**

In example *a*, verification is achieved by numerical methods. In example *b*, verification is determined empirically. The main idea here is that students can verify a procedure independently rather then rely on the teacher or the textbook for verification. Of particular importance is the situation in which a student discovers that an answer does not check out and hence reexamines the execution of the original procedure. Assessment should focus on whether students know how to verify a result and whether they can complete the verification process.

3. *Generate new procedures or modify familiar ones.*

In figure 9.1, use only a compass to find a point *X* such that *PX* (if drawn) would be parallel to line *m*. Describe and justify your procedure.

Fig. 9.1

A ruler should *not* be used to solve this problem. Assessment should focus on whether the construction is complete and accurate, on the student's description of what was done, and on the rationale for the procedure. The descriptions and explanations can be written or oral.

4. *Appreciate the nature and role of procedures in mathematics.*

A single item or task offers insufficient evidence to assess a student's appreciation of the nature and role of procedures in mathematics. A valid assessment must occur over time and take into account students' remarks and actions in a variety of mathematical activities or tasks that call for the use of well-known procedures or the generation of new ones. For this aspect of procedural knowledge to be realized, it is essential that the instructional program provide opportunities for students to generate procedures. Such opportunities should dispel the belief that procedures are predetermined sequences of steps handed down by some authority (e.g., the teacher or the textbook). Important questions to be considered in assessing this aspect of students' procedural knowledge include the following:

- Do students see that procedures are generated for a purpose or to meet a specific need?

- Do students value participation in the generation or extension of procedures?
- When students cannot recall a particular procedure, do they attempt to reconstruct the procedure or generate a new one, rather than seek help in recalling the forgotten procedure?
- Do students see that alternative procedures can meet the same need?
- Do they judge the relative merits of alternative procedures on the basis of their efficiency?

Furthermore, when a new procedure is introduced, the following questions should be assessed:

- Do students attempt to make sense of the sequence in which the steps are carried out?
- Do they question the logic in the sequence of steps?
- Do they question why a given procedure produces the desired results?
- Do they try to verify their results?

These behaviors can be indicative of students' understanding of the nature and role of procedures.

STANDARD 10: MATHEMATICAL DISPOSITION

The assessment of students' mathematical disposition should seek information about their—

- ***confidence in using mathematics to solve problems, to communicate ideas, and to reason;***
- ***flexibility in exploring mathematical ideas and trying alternative methods in solving problems;***
- ***willingness to persevere in mathematical tasks;***
- ***interest, curiosity, and inventiveness in doing mathematics;***
- ***inclination to monitor and reflect on their own thinking and performance;***
- ***valuing of the application of mathematics to situations arising in other disciplines and everyday experiences;***
- ***appreciation of the role of mathematics in our culture and its value as a tool and as a language.***

Focus

Learning mathematics extends beyond learning concepts, procedures, and their applications. It also includes developing a disposition toward mathematics and seeing mathematics as a powerful way for looking at situations. *Disposition* refers not simply to attitudes but to a tendency to think and to act in positive ways. Students' mathematical dispositions are manifested in the way they approach tasks—whether with confidence, willingness to explore alternatives, perseverance, and interest—and in their tendency to reflect on their own thinking. The assessment of mathematical knowledge includes evaluations of these indicators and students' appreciation of the role and value of mathematics.

This kind of information is best collected through informal observation of students as they participate in class discussions, attempt to solve problems, and work on various assignments individually or in groups. Such assessment procedures as attitude questionnaires fail to capture the full range of perceptions and beliefs that underlie students' dispositions.

Discussion

From their first encounter with shapes or numbers, children begin to form a conception of mathematics. Teachers implicitly provide information and structure experiences that form the basis of students' beliefs about mathematics. These beliefs exert a powerful influence on students' evaluation of their own ability, on their willingness to engage in mathematical tasks, and on their ultimate mathematical disposition.

Mathematical disposition is much more than a liking for mathematics. Students might like mathematics but not display the kinds of attitudes and thoughts identified by this standard. For example, students might like mathematics yet believe that problem solving is always finding one correct answer using the right way. These beliefs, in turn, influence their actions when they are faced with solving a problem. Although such students have a positive attitude toward mathematics, they are not exhibiting essential aspects of what we have termed *mathematical disposition*.

The assessment of students' dispositions requires information about their thinking and actions in a wide variety of situations and should consider all aspects of disposition and the degree to which they are exhibited. Disposition has many components, each of which a particular student exhibits to a greater or lesser extent. For example, a student might be very willing to try alternative methods of solving problems but be less inclined to reflect on the solutions. Another student might be fairly uninterested in routine exercises and yet work diligently to solve nonroutine problems. An adequate assessment of these students' disposition requires information on their willingness to engage in all aspects of solving problems, including learning through problem solving.

In the classroom students' dispositions are continuously reflected in how they ask and answer questions, work on problems, and approach learning new mathematics. As a result, teachers are in an excellent position to gather useful information for assessing disposition. In addition, teachers benefit from this assessment because it provides information for instructional planning. If an assessment indicates that most students in a class rarely attempt problems independently and frequently ask to be shown the solution method, a teacher can choose to reexamine classroom instruction to evaluate whether students are being encouraged to solve problems and develop alternative solutions. In short, the assessment of students' dispositions provides information about changes needed in instructional activities and classroom environments to promote the development of students' mathematical dispositions.

Because evidence of students' mathematical dispositions is apparent in every aspect of their mathematical activities, observation is a primary method of assessment. When presented with a problem, particularly one in a new and unfamiliar context, a student exhibits his or her mathematical disposition in a willingness to change strategies, reflect and analyze, and persist until a solution is identified. Students' disposition toward mathematics can be observed in class discussion. How willing are students to explain their point of view and defend that explanation? How tolerant are they of nontraditional procedures or solutions? Are they curious? Are they willing to ask, "What if . . .?" What kind of questions do they ask?

Although observation is the most obvious way of obtaining such information, students' written work, such as extended projects, homework assignments, and journals, as well as their oral presentations, offer valuable information about their mathematical dispositions. Such projects as individual or small-group presentations of problem solutions or proofs of theorems can serve as evidence of students' willingness to persevere at mathematical tasks and test alternative methods in solving problems.

Grades K—4

The assessment of mathematical disposition of students in the primary grades should focus on the number and quality of students' experiences. An inventory like the one in figure 10.1, when kept for a class or for individual students, can be used as a record of their experiences. This record can be used for determining if a full range of experiences has been offered and for reporting to parents, guardians, and other teachers the experiences of the students.

Inventory of Mathematical Disposition Experiences

What students have experienced:	Date and Activity	
1. Confidence in using mathematics	10/29 Correctly solved all problems assigned.	1/19 Actively worked as part of small group that solved a problem.
2. Flexibility in doing mathematics	11/2 Generated several ways of solving an addition problem.	4/8 Students challenged each other on solution methods.
3. Persevering at mathematical tasks	4/29 Worked all day on collecting and displaying data—favorite ice cream.	2/5 Kept working on different ways of making change for 50¢—all ways found.
4. Curiosity in doing mathematics	11/10 Solved a "what if" question, expressing answer in own words.	4/7 In small group generated own units for measuring room.
5. Reflecting on their own thinking	Every day students explain their thinking after working on a problem.	
6. Valuing applications of mathematics	2/23 All students brought in pictures for math applications bulletin board.	5/24 Field trip to science museum to see how mathematics is used.
7. Appreciating role of mathematics	10/15 Brought in newspaper articles that used mathematical terms.	1/29 Appreciated place-value system by finding sums using Roman numerals.

Fig. 10.1

Grades 5–8

In the middle grades, students' mathematical dispositions become more apparent in their daily work. A checklist of actions that demonstrate mathematical disposition can be used to help record students' progress. Such a checklist can be completed by a teacher while focusing on five students during a class period. Over six weeks, each student's mathematical disposition can be evaluated at least once. A sample checklist is given in figure 10.2.

Grades 9–12

Essays and research papers are particularly useful for both developing and assessing students' appreciation of the role of mathematics in our culture. Judgments about the quality of the work should be based on its conceptual merit, thoroughness, and originality. Just as the assessment of writing includes a consideration of grammar, spelling, fluency of expression, force of ideas, and consistency of logic and narration, so mathematical assessment should include considerations of quality as well as accuracy. An innovative solution, a well-reasoned argument, and a willingness to go beyond the constraints of the task are of equal importance in obtaining a correct solution and should be acknowledged. Sample topics are the following:

- The role of geometry in our society today
- Mathematics and computers

Mathematical Disposition Checklist

Date: Name / Action observed	Student 1	Student 2	Student 3	Student 4	Student 5
Confidence: **Initiates questions.**	✓		✓	✓	
Is sure answers will be found.		✓	✓	✓	
Helps others with problems.	✓			✓	
Other/note:					Challenged the solution of another student
Flexibility: **Solves problems in more than one way.**	✓	✓			
Changes opinion when given a convincing argument.			✓	✓	
Other/note:		created new problem by changing cond.			

Fig. 10.2

- What if standard units of measures had not been created
- Changes in the applications of statistics in sports

No one of the preceding considerations across all grades will yield a definitive picture of a student's disposition toward mathematics. However, extended observations of students' efforts and interactions in a variety of mathematical contexts can give teachers the feedback and information necessary to adjust their instructional methods and encourage students' progress in attaining intellectual autonomy.

PROGRAM EVALUATION

The Program Evaluation Standards are an integral component of the overall vision of evaluation's primary role in guiding change. The vision includes program change so that all the pieces will fit together. This change will provide the foundation in grades K–4 for students to know mathematics through doing mathematics, will develop and expand the ideas in grades 5–8, and will lead to self-directed learning in grades 9–12. Evaluation can help determine a mathematics program's status in relation to the Curriculum Standards and ensure that the pieces fit together. It can indicate the steps that need to be taken so that a program aligns with the *Standards*. In a certain sense, the Program Evaluation Standards are a guide to creating a program that meets the challenge of the *Standards*.

One purpose of program evaluation is to obtain relevant and useful information for making decisions about curriculum and instruction. The assessment of students' capabilities, as described in the previous ten evaluation standards, is one means of collecting information about a program. The principles of general assessment apply as much to program evaluation as to student assessment. But there are other indicators of program quality. Does the curriculum include the range of topics and emphases described in the *Standards*? Is the program articulated from kindergarten through grade 12 so that students' knowledge and experiences build continually? Does prior mathematics study provide the conceptual underpinnings for additional study and for its application to real-world problems? Does instruction actively involve students in the investigation, exploration, and development of mathematical ideas and the relationships among them? Are calculators and computers used appropriately? Do all students have full access to the program? Do students know mathematics as an integrated whole?

Any program can be implemented by degrees. At one level, the language of the new program can be adopted while day-to-day instruction remains unchanged. At another level, minor changes in structure can be made by inserting a new unit into a course or slightly modifying the scope and sequence. At a third level, deep structural changes can be made that include altering how people think about a program, how mathematics is presented, and how students come to know mathematics. The *Standards* speaks to this third level of change. It is the role of program evaluation to facilitate this deep structural change.

◆ ◆ ◆ ◆ ◆ ◆ ◆ ◆

STANDARD 11: INDICATORS FOR PROGRAM EVALUATION

Indicators of a mathematics program's consistency with the* Standards *should include—

- ***student outcomes;***
- ***program expectations and support;***
- ***equity for all students;***
- ***curriculum review and change.***

In addition, indicators of the program's match to the* Standards *should be collected in the areas of curriculum, instructional resources, and forms of instruction. These are discussed in Evaluation Standards 12 and 13.

Focus

The central theme of the *Standards* is that knowing mathematics is doing mathematics. Through solving problems, reasoning, communicating, investigating, and exploring, students will know mathematics. Inherent in this theme is that any evaluation of a program's match to the *Standards* must consider what mathematics students know; how they learn mathematics; and the curriculum, means of instruction, and expectations of those who influence the program. In addition, an evaluation should consider any barriers that prevent students from attaining the full benefits of the program and what can be done to eliminate such barriers. Finally, evaluation should consider the dynamic nature of the program, including its process for self-monitoring and for making necessary adjustments and changes.

Discussion

A primary purpose of any evaluation, using the *Standards* as a criterion, is to obtain information and suggestions about how a given program can more fully incorporate the spirit—as well as the letter—of the *Standards.* An evaluation should provide evidence that a program does or does not match the *Standards.* In addition, an evaluation should gather information on how a program can be changed to better meet the *Standards.* Thus, an evaluation should collect information on a range of indicators that signify a match between the mathematics program and the *Standards* and that help explain what needs to be done differently.

Student Outcomes. Students in a mathematics program that is consistent with the *Standards* will become knowledgeable about a variety of topics and the relationships among them and will develop a positive disposition toward mathematics. The previous Evaluation Standards have described the means and possible indicators for assessing students' knowledge and disposition. This discussion focuses on the assessment of mathematics programs as a whole.

A valid evaluation of a program's alignment or match with the *Standards* depends on a very comprehensive assessment of students' mathematical knowledge. A great many topics and processes should be evaluated using a variety of methods, such as observations of students doing mathematics, performance and oral tasks, and written tests. To achieve adequate coverage, for example, an assessment of student outcomes for one grade, such as the eighth grade, might require collecting information on at least seven indicators, one for each of the student-assessment standards. Such an assessment would include a number of tasks, reaching a possible total of 300 tasks or test questions. Obviously, even a comprehensive assessment cannot require each student to answer all questions, but this is not necessary for a valid or purposeful program evaluation. A procedure such as multiple-matrix sampling should be used to reduce substantially both the number of students and the amount of time any single student would need to be involved. The use of such a procedure addresses the issue of too much testing being done in schools.

Indicators of student outcomes should be aligned with the Curriculum Standards. Norm-referenced tests, for example, reflect too limited a range of content to measure adequately the *Standards*' expectations of what students should know. As such, norm-referenced test scores, as the sole indicator of student outcomes, are inappropriate for comparing a mathematics program with the *Standards.*

The evaluation of student outcomes of a program should give some attention to long-term effects. These can be considered in longitudinal studies of the effects of the program on the lives of its graduates one, two, or five years after they have graduated. Such studies should try to establish the effects of the program in helping the graduates reach their goals and meet the challenges they face after finishing the program.

Program Expectations and Support. An adequate support system is a prerequisite to bringing a mathematics program into alignment with the *Standards.* Such a system is based on the view that a variety of instructional activities are needed and that all students must know the mathematics described in the *Standards.* If such a view is prevalent throughout the school district and shared by most decision makers—including mathematics supervisors, district administrators, and school board members—the likelihood of a program coinciding with the *Standards* is increased. Indications of a strong support system are the existence of staff-development programs and the provisions made to allow students sufficient time to take an inquiry approach in learning mathematics. Evaluation should offer some evidence on whether the necessary expectations and support exist.

Program evaluation also should consider the degree to which parents are interested, knowledgeable, and involved in the mathematics education of their children. Do parents take an interest in their children's progress in learning mathematics? Do parents monitor the time students spend on doing mathematics homework compared with other activities, such as watching television? What ways can parental involvement be increased in light of the *Standards*' view of mathematics?

Equal Access. A critical component of any mathematics program, particularly one that strives to fulfill the spirit of the *Standards,* is the equal access that all students have to take advantage of the full benefits of the program. The official position of the NCTM is that "all students, regardless of their language or cultural background, must have access to the full range of mathematics courses offered. Their patterns of enrollment and achievement should not differ substantially from those of the total student population" (NCTM 1987). Program evaluations should include indicators that the mathematics program is meeting these essential criteria. Enrollment figures by gender, race, language, and cultural background should be maintained for all mathematics courses. As unacceptable patterns emerge, an evaluation should identify the barriers creating the situation and recommend action. All students should have equal access to the full range of mathematics courses, and this should be continually monitored and included as a part of the ongoing program review.

Curriculum Review and Change. Just as the world is changing rapidly and new technologies are developing continually, so must mathematics programs evolve and grow. The vision articulated in the *Standards* is that all children will develop the mathematics they need to function and succeed in a world in which the ability to think about and solve problems has become increasingly important. It is obvious, however, that the *Standards* cannot foresee all the changes in calculators, computers, and software or other advances that might affect what mathematics students will need to know in the future. One function of evaluation is to determine whether a given program has established a self-monitoring process by which it keeps current with the dynamic nature of mathematics.

STANDARD 12: CURRICULUM AND INSTRUCTIONAL RESOURCES

In an evaluation of a mathematics program's consistency with the Curriculum Standards, the examination of curriculum and instructional resources should focus on—

- ***goals, objectives, and mathematical content;***
- ***relative emphases of various topics and processes and their relationships;***
- ***instructional approaches and activities;***
- ***articulation across grades;***
- ***assessment methods and instruments;***
- ***availability of technological tools and support materials.***

Focus

What mathematics is taught in the classroom and *how* it is taught are strongly influenced by a district's curriculum and adopted materials. Therefore, a necessary first step in determining the extent to which a mathematics program meets the *Standards* is examining the curriculum and the materials used in the program.

The *Standards* offers a vision of, and a direction for, a mathematics curriculum but does not constitute a curriculum in itself. If a mathematics program is to be consistent with the *Standards*, its goals, objectives, mathematical content, and topic emphases should be compatible with the *Standards'* vision and intent. Likewise, the instructional approaches, materials, and activities specified in the curriculum should reflect the *Standards'* recommendations and be articulated across grade levels. In addition, the assessment methods and instruments should measure the student outcomes specified in the *Standards*.

Discussion

Achieving agreement between a curriculum and the *Standards* is a necessary first step in having a mathematics program that reflects the vision expressed by the *Standards*. However, such agreement is not a guarantee that the curriculum, as implemented, will fully comply with the *Standards*. There is a difference between a curriculum as specified in a plan or textbook and a curriculum as implemented in the classroom. A curriculum specifies goals, topics, sequences, instructional activities, and assessment methods and instruments. An implemented curriculum is what actually happens in the classroom.

A deep, thorough analysis is necessary to determine the extent to which a curriculum and its materials are compatible with the *Standards*. The *Standards* offers a framework for curriculum development but not a scope and sequence. Simply checking topics on a scope-and-sequence chart is insufficient to determine the extent to which a curriculum and its materials are compatible with the *Standards*. A comparative analysis must provide qualitative documentation of the degree of consistency between the *Standards* and the curriculum. Such results can then be used to make decisions about the adoption of materials and how the curriculum needs to be modified to be more consistent with the *Standards*.

Goals, Objectives, and Content. The curriculum should include the major goals of developing students' problem-solving, reasoning, and communication abilities. In line with the *Standards*, however, these processes should not be listed as topics or separate strands to be taught in isolation. Rather, the curriculum and the adopted materials must provide a means of, and directions for, ensuring that these processes are integrated across topics. A superficial analysis will not suffice; for example, to determine how well problem solving as a curricular goal is reflected in the materials, it is necessary to look beyond such surface characteristics as the number of problem-solving lessons in a chapter. If problem solving is limited to a few lessons in each chapter and not integrated into the development of content topics, then the materials fall short of the *Standards'* intent.

The evaluation of instructional materials should take into account the quality of activities, their intended use in instruction, and their frequency of use. For example, textbook series commonly claim that problem solving is integrated into their programs. This claim is based, in part, on the common use of problems as "lesson openers." Yet, the lesson development that follows is, at best, remotely connected to the problem itself and is often didactic and prescriptive in nature. Such lessons fail to capture the spirit of problem solving and can convey an incorrect message to teachers and students about what problem solving really means.

Emphases and Relationships. The Curriculum Standards prescribe that emphasis be given to the meaningful development of concepts, to the interrelationships or connections among topics, and to the application of mathematics to the solution of realistic problems. These emphases should be explicitly and clearly stated in curricular and instructional materials and reflected in the development of lessons and activities following each lesson. For example, the *Standards* points out the importance of achieving an appropriate balance between concept development and computational proficiency. An inordinate emphasis on skill proficiency at the expense of a strong conceptual framework to support such skills renders materials inappropriate.

When feasible, curricular materials should develop new topics or ideas as natural extensions or variations of ideas students already know, thus making connections among topics explicit. Alternative representations and meanings of a mathematical idea should be featured in lessons as well as in practice activities. In addition, materials should allow ample opportunities for students to apply the mathematics they have learned in realistic and meaningful situations.

Instructional Approaches and Activities. When materials are evaluated for use in the classroom, a primary consideration is whether they promote students' active involvement in learning mathematics. Approaches and activities that call for the investigation and exploration of ideas, problem solving, conjecturing and testing of conjectures, and verification should be integrated through all the lessons rather than included as occasional isolated features at the end of a chapter or unit. Furthermore, such activities should be appropriate as student assignments in that they offer directions and guidance on how to use such approaches and activities effectively.

♦ ♦ ♦ ♦ ♦ ♦ ♦ ♦

Articulation across Grades. The curriculum and materials should provide for the natural and logical development of mathematical topics across grades. Topic development should follow a progression from building a foundation of conceptual understanding, to extending concepts and procedures, to formalizing and abstracting ideas. A program that is well articulated will include directions for introducing, extending, and formalizing what has been taught before, with emphasis on the connections among related topics. The materials should also give teachers of different grades ways to coordinate the development of topics across the curriculum.

Assessment Methods and Techniques. Comparison of a curriculum with the *Standards* should include a careful study of what students are expected to know. To achieve agreement between the two, the assessment methods and instruments specified in the curriculum must be aligned with the student outcomes set forth in the *Standards* (see Evaluation Standard 1 on alignment). Tasks should be broad in scope and tap the extent to which students have integrated their knowledge of concepts, procedures, and processes. In addition, assessment instruments should reflect the relative emphases of various topics and processes as specified in the *Standards*.

Technological Tools and Resources. The mathematics classroom envisioned in the *Standards* is one in which calculators, computers, courseware, and manipulative materials are readily available and regularly used in instruction. Although no rigid criteria exist for judging what constitutes adequate resources and equipment, a program evaluation should include information about the resources available so that funds can be allocated properly.

STANDARD 13: INSTRUCTION

In an evaluation of a mathematics program's consistency with the Curriculum Standards, instruction and the environment in which it takes place should be examined, with special attention to—

- ***mathematical content and its treatment;***
- ***relative emphases assigned to various topics and processes and the relationships among them;***
- ***opportunities to learn;***
- ***instructional resources and classroom climate;***
- ***assessment methods and instruments;***
- ***the articulation of instruction across grades.***

Focus

How mathematics is taught is just as important as *what* is taught. Students' ability to reason, solve problems, and use mathematics to communicate their ideas will develop only if they actively and frequently engage in these processes. Whether students come to view mathematics as an integrated whole instead of a fragmented collection of arbitrary topics and whether they ultimately come to value mathematics will depend largely on how the subject is taught. Thus, the evaluation of a mathematics program must include an analysis of instruction as well as content.

Discussion

An evaluation of how mathematics is taught must consider how well the program is being implemented and how instruction can be modified to better meet its goals and objectives. Such an analysis requires the collection from a representative sample of teachers of various types of information, including, but not limited to, sustained classroom observations, interviews with teachers, teachers' written self-reports, questionnaires, and summaries of peer observations.

Each of the following sections offers guidance for gathering information to evaluate instruction. Both quantitative and qualitative information must be obtained.

How Content Is Treated. An evaluation of instruction should look first at the accuracy of the mathematics that is taught. In addition, an evaluation should determine whether that mathematical content is treated in a manner that is sensitive to the developmental level and mathematical maturity of the students. Instruction that is sensitive to the students' developmental level is set in realistic contexts, incorporates students' experiences, uses language that is suited to the maturity of students, allows sufficient time for students to construct meanings by exploring and investigating mathematical ideas, and offers students opportunities to discuss their ideas.

◆ ◆ ◆ ◆ ◆ ◆ ◆ ◆

The *Standards* specifies reasonable expectations for students and offers suggestions for the treatment of topics at various grade levels. Students' needs and characteristics were primary considerations as the *Standards* was developed. Nonetheless, because the needs and characteristics of students vary greatly even at a single grade level, it is important to take these variables into account when examining instruction.

The *Standards* advocates students' active involvement in learning, a stance that has important implications for the way content is to be treated during instruction. Rather than a routine presentation of mathematical ideas in a polished, finished form for students to assimilate, instruction should provide frequent opportunities for students to generate, discuss, test, and apply mathematical ideas and verify their findings.

Such instructional approaches also have implications for the role of the teacher during instruction. When using a problem-solving approach in developing an idea, for example, teachers must encourage students to guess courageously. Teachers must be willing to entertain suggestions from students and suspend judgment about their ideas. Teachers should help students evaluate one another's suggestions and critically reflect on them by anticipating objections and consequences. Clearly, these activities require the teacher to assume a role very different from that of a directive authority. Classroom observations for the purpose of evaluating instruction should focus on the role of the teacher and the appropriateness of that role to the content and activities.

Topic Emphases and Relationships. The *Standards* proposes that instruction should emphasize interrelationships among mathematical ideas. Classroom observations should gather information about whether mathematics is portrayed as an integrated body of logically related topics as opposed to a collection of arbitrary rules that students must memorize. For example, the K–4 standards suggest ways in which development and extension of number concepts and operations can be integrated into the treatment of other topics, such as measurement, geometry, patterns, and graphs. The extent to which new ideas are presented as natural or logical extensions of ideas the students have already encountered should be a focal point of instructional evaluation.

The *Standards* calls for an instructional emphasis on building strong conceptual frameworks on which to base the development of skills. It also emphasizes the importance of multiple representations of a mathematical idea and the translation of an idea from one representational system to another. A documentation of the extent to which these and other instructional emphases in the *Standards* are implemented should be a major part of any serious evaluation.

Classroom observations should also document the extent to which instruction emphasizes the relationships among the various branches of mathematics and between mathematics and other areas of the school curriculum.

Opportunity to Learn. If instruction is to result in the student outcomes specified in the *Standards*, students need to have sufficient opportunities to learn the specified content. Thus, program evaluations must consider the amount of time *actually* devoted to mathematics instruction; an hour of mathematics each day at all grade levels is a reasonable expectation. The frequency of interruptions caused by school assemblies and other school projects must be considered in the evaluation to determine the actual time.

Further, evaluation should focus on the attention given at each grade level to the various branches of mathematics, such as geometry, measurement, statistics, probability, algebra, discrete mathematics, and calculus in light of the recommendations in the *Standards.*

Similarly, if students are to become competent in problem solving, reasoning, and communicating mathematically, instruction must allow them opportunities to engage actively and frequently in these processes. Classroom observations for evaluating instruction must pay special attention to the frequency and quality of instructional activities that afford students such opportunities.

Instructional Resources and Classroom Climate. The evaluation of instruction should determine the extent to which the classroom environment is conducive to the attainment of program goals and student outcomes. One indicator would be the availability and use of instructional resources, such as computers, calculators, courseware, and manipulative materials. In addition, evaluation should determine whether an intellectual "climate" exists, in which students' curiosity, openness, and spontaneity are encouraged.

Uses of Classroom Assessment Methods. The methods, instruments, and tasks used to assess students' learning should be consistent with the content taught and the emphases placed on various topics and processes. Evaluation Standards 1–10 present guidelines for assessing students' outcomes as well as for judging the adequacy of the instruments and tasks used in that assessment.

Articulation of Instruction. An evaluation should examine the extent to which the implemented program is articulated across grades. The degree of consistency between the *Standards* and instruction should be determined on the basis of data obtained from a number of classrooms in a given school. It is insufficient to determine that one or two classrooms exhibit this consistency. Within a given school, evaluation must obtain information about the extent to which teachers have opportunities to coordinate their instruction with the *Standards* across grades.

STANDARD 14: EVALUATION TEAM

Program evaluations should be planned and conducted by—

- ***individuals with expertise and training in mathematics education;***
- ***individuals with expertise and training in program evaluation;***
- ***individuals who make decisions about the mathematics program;***
- ***users of the information from the evaluation.***

Focus

An evaluation of a mathematics program should be planned and directed by a team whose members are competent, enthusiastic, and committed. This team should be composed of those who have the greatest interest in the results of the evaluation—the stakeholders. These include the program decision makers and the users of the information that will be produced. In a mathematics program, these participants are teachers, mathematics supervisors, school and district administrators, school board members, and members of the community. Here teachers are considered both decision makers and users. Teachers can and do make critical curriculum decisions and should participate in any program evaluation. In addition, mathematics educators should be involved in interpreting the spirit and the letter of the *Standards*. The team should also include a program evaluator with some knowledge of evaluation procedures that coincide with Program Evaluation Standards 11–13.

Such an evaluation team includes people who have the greatest need for information about the program and who care most about what happens to it. An evaluation conducted by such a team has at least one key advantage over any conducted by a single evaluator or small group: A panel of participants with vested interest in the program is more apt to generate the important questions to be addressed in the evaluation, thereby increasing the likelihood that the effort will produce useful and meaningful information.

Discussion

An evaluation often is threatening, especially when it is conducted by "outside" experts. The results of such an evaluation might be regarded with suspicion and therefore might be of limited use. When individuals involved in a program also participate in its evaluation, they have an enhanced sense of "ownership." They tend to view the evaluation conclusions as more valid and more useful in decision making. Some observers might question the credibility of evaluation findings from efforts directed largely by insiders. School districts must be sensitive to this issue and consider it in light of local conditions and circumstances. An outsider's view might be warranted in some situations, or someone with more objectivity or different ideas might be needed. However, the team approach to evaluation places a higher value on producing a useful and valid evaluation than on meeting all the requirements of a less subjective form.

The program evaluator and the mathematics educator are responsible for bringing their expertise to the evaluation. A program evaluator should

♦ ♦ ♦ ♦ ♦ ♦ ♦ ♦

know how to phrase questions and develop procedures for collecting information. A mathematics educator will bring a knowledge of mathematics and instruction to the process. Such knowledge is helpful in identifying how the full meaning of the *Standards* relates to the existing program. Some districts might need to bring in outside experts if they cannot be identified from within.

Because of the primary role of teachers in achieving the vision of the *Standards,* their participation is crucial to an effective and useful evaluation. Since the implementation of the program can vary by grade, it is important that the evaluation team include teachers from different grade levels. At a minimum, representatives from each of the grade-level ranges established in the *Standards* should be included; this requirement applies to teams evaluating programs in elementary, middle, or secondary schools (i.e., for districts that are K–4, 5–8, or 9–12), as well as for K–12 programs. For example, the evaluation team for the mathematics program of an elementary school district should include a teacher from the neighboring middle school district to ensure articulation across programs.

School and district administrators who help create and implement the curriculum are particularly important members of any evaluation team. In addition to their knowledge of the curriculum, they, along with teachers, can provide insights into methods of implementation and related problems, as well as up-to-date information about available support.

In addition, the team may include other individuals who are interested or involved in making decisions regarding the outcomes of the mathematics program, such as parents, school board members, and representatives of the business and work communities. Members from institutions or organizations other than schools are useful because they have different viewpoints and can act as interpreters between the school and the community.

The membership and function of an evaluation team can vary according to existing conditions, availability of local expertise, and resources. In all situations, however, it is crucial that a team of interested people participate in the evaluation process rather than leave the task to one or two individuals. Every team member need not be involved in every part of the evaluation; a team may at times function as a body of subgroups, each assigned a particular function that best taps its members' expertise.

As described in Evaluation Standard 13, classroom observations are essential to determine how actual instruction compares with the *Standards'* vision. An adequate evaluation of instruction requires the participation of someone knowledgeable about the mathematics being taught and the goals and approaches envisioned by the *Standards.* If students are using manipulatives, the observer should know what mathematics is being modeled, whether the manipulatives are appropriate, and whether they are being used correctly.

A thorough evaluation of a mathematics program is a major task. It requires considerable time for both planning and implementation. A district must provide members of the evaluation team with the resources and time necessary to assess with care all aspects of the program: the curriculum, the implemented curriculum, and student outcomes.

NEXT STEPS

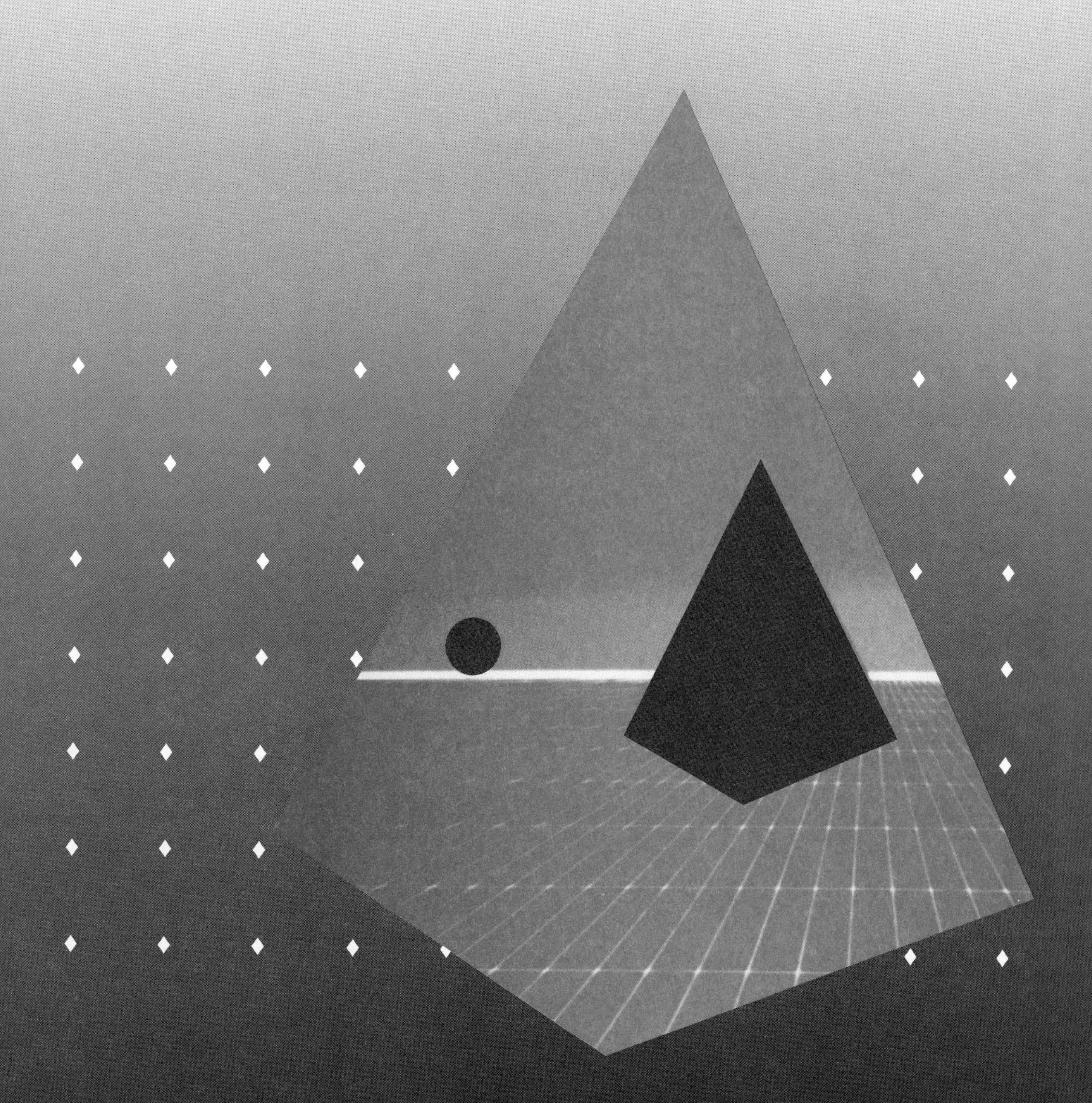

♦ ♦ ♦ ♦ ♦ ♦ ♦ ♦

NEXT STEPS

Changing School Mathematics

Deciding on the content of school mathematics is the initial step in the necessary change process. So that the next steps proceed in harmony with the *Standards*, both the nature of the changes needed and the strategy implied for change should be understood. Given the overwhelmingly positive response to the Working Draft of the *Standards*, the commission members and writers are convinced that hundreds of teachers and other mathematics educators are eager to change school mathematics. In fact, we are optimistic that such changes can and will be accomplished.

The Nature of Change. In any field, when systems are not working, those involved in them must decide whether the problem is the result of a lapse in quality control, a design flaw, or a combination of the two. Quality-control solutions improve the efficiency and effectiveness of what is being done without disturbing its basic features. Design solutions fundamentally alter the organization of the systems themselves. Both of these solutions are applicable to the subject of school reform. Quality-control changes in education, for example, have included "recruiting better teachers and administrators, raising salaries, allocating resources more equitably, selecting better textbooks, adding (or deleting) content or course work, scheduling people and activities more efficiently, and introducing new versions of evaluation and training" (Cuban 1988, p. 342). However, design changes in education "introduce new goals, structures, and roles that transform familiar ways of doing things into new ways of solving persistent problems" (Cuban 1988, p. 342).

Given this distinction about strategies for change, it should be obvious that we see the *Standards* as an initial step in a design-change process. The changes advocated in the *Standards* should lead to a fundamental restructuring of the mathematics curriculum and instruction. However, the redesign we envision cannot be done in a mechanical fashion. Instead, teachers and other educators must come to a consensus and work collaboratively to bring about the changes needed. The remaining steps, and there are many, in the redesign of school mathematics are based on the following strategy for change.

The Professional Development Change Strategy. In the past most educational changes have been approached through a "top down" managerial strategy borrowed from industry. Managers and experts design new parts or procedures and then train workers to use them. The change strategy being advocated here, however, is based on professional development rather than administrative directives.

Professional development implies the direct involvement of professional organizations and their members. The vision of the mathematics curriculum and the goals in the *Standards* are offered by NCTM without a prescription for achieving them. The approach taken is one that will empower teachers and other educators, through their professional organizations, to make the changes. Thus, it is the responsibility of all those involved and interested in school mathematics to implement reform.

Necessary Next Steps

The next steps toward change should not be considered as linear or exhaustive but rather as steps along many paths headed in the same direction. Professionals in different areas will follow different paths to redesign components for a new system of school mathematics. Some of those paths follow:

Curriculum Development. The *Standards* is a framework for curriculum development. However, it contains neither a scope-and-sequence chart nor a listing of topics by specific grade level. This is deliberate; a coherent network of relationships exists among the identified topics, and multiple paths are available throughout this network. What we have done is to identify the primary elements, or nodes, of the network to be included in a quality mathematics curriculum. One possible next step is for teachers and mathematics educators to develop curricula based on the standards.

For example, the secondary school mathematics curriculum has typically been separated into courses with a specific subject orientation (e.g., algebra, geometry, statistics). This sequence provides teachers and students with a single-course focus. We now challenge educators to integrate mathematics topics across courses so that students can view major mathematical ideas from more than one perspective and bring interrelated ideas to bear on new topics or problems. Similarly, texts in grades K–8 often include a variety of mathematical topics but primarily stress arithmetic. We favor instead a truly integrated curricular organization in all grades to permit students to develop mathematical power more readily and to allow the necessary flexibility over time to incorporate the content of these standards. This integration is intended both among mathematical topics and with their use in other subjects as expressed in the "Connections" standards at each level.

Textbooks and Other Materials. We are aware that the curriculum in many schools is geared to textbooks, and we expect the standards to be used as criteria for measuring text content. However, we do not believe the standards can be met simply by altering current texts (e.g., appending a genuine problem at the end of each chapter or inserting a chapter on statistics). Those are quality-control solutions to a problem that demands a design solution. Nor do we believe that textbooks should drive instruction. Rather, other materials that support the standards, such as manipulatives and courseware, must be developed, in addition to new textbooks. As an initial step on this path, NCTM is developing addenda to the standards that will contain many exemplary instructional activities.

Tests. Tests have an influence on what actually gets taught in a classroom, especially in urban areas where teachers know that the test results will be used, rightly or wrongly, as an evaluation of *them.* New tests must be developed to assess problem solving, reasoning, and so on, in a valid way to ensure that the mathematics intended in the *Standards* is taught in *all* classrooms. Such new testing strategies must be applied not only to standardized tests but to state testing programs as well. Finally, until tests provide for the appropriate use of calculators, many teachers will continue to prohibit their use in the classroom. Without changes in how mathematics is assessed, the vision of the mathematics curriculum described in the standards will not be implemented in classrooms, regardless of how texts or local curricula change.

Instruction. The spirit and vision of the *Standards* cannot be achieved if instruction is inconsistent with its underlying philosophy. Specifying the content for a quality mathematics program is impossible without addressing the accompanying instructional conditions. Thus, the elaboration of each standard deliberately contains implications for instruction and includes expectations about teachers' actions, such as the use of a variety of sequences, grouping procedures, instructional strategies, and techniques for evaluation. However, the *Standards* was not developed as

◆ ◆ ◆ ◆ ◆ ◆ ◆ ◆

a set of instructional standards. NCTM has formed several task forces to develop materials to assist with the delivery of instruction.

Teacher In-Service Programs. Although we are confident that many teachers are now ready to teach the kind of mathematics program outlined in the *Standards*, many others will need and demand additional training or refresher courses. These programs must be developed in collaboration with the teachers. They must include mechanisms for sustained collegial interaction, links between staff development and classroom practice, and the participation of administrators to ensure support for the proposed changes. We challenge states, provinces, and school districts to use their in-service resources to help teachers in this manner. Again, NCTM has created an implementation committee to coordinate and orchestrate the development of in-service materials and programs.

Teacher Education. Prospective teachers must be taught in a manner similar to how they are to teach—by exploring, conjecturing, communicating, reasoning, and so forth. Thus, colleges of education and mathematical sciences departments should reconsider their teacher preparation programs in light of these curriculum and evaluation criteria. Teacher education programs now being redesigned to follow the recommendations of such groups as the Holmes Group and the Carnegie Commission must be compatible with the *Standards*. To implement these standards, all teachers need an understanding of both the historical development and current applications of mathematics. Furthermore, they should be familiar with the power of technology. The NCTM is developing professional standards for teaching mathematics.

Technology. Throughout each standard, we have assumed that appropriate technology is available for classroom instruction. Calculators, computers, courseware, and manipulative materials are necessary for good mathematics instruction; the teacher can no longer rely solely on a chalkboard, chalk, paper, pencils, and a text. Criteria for measuring the adequacy of materials and equipment must be developed. Note, however, that simply providing teachers with these materials will not produce a new program; teachers must also know how to integrate this technology into a quality mathematics program.

Students with Different Needs and Interests. The consequences of dealing with students with different talents, achievements, and interests have led to such practices as grouping and tracking and to special programs for gifted or handicapped students who need and deserve special attention. However, we believe that *all* students can benefit from an opportunity to study the core curriculum specified in the *Standards*. This can be accomplished by expanding and enriching the curriculum to meet the needs of each individual student, including the gifted and those of lesser capabilities and interests. We challenge teachers and other educators to develop and experiment with course outlines and grouping patterns to present the mathematics in the *Standards* in a meaningful, productive way.

Equity. As a pluralistic, democratic society, we cannot continue to discourage women and minority students from the study of mathematics. We believe that current tracking procedures often are inequitable, and we challenge all to develop instructional activities and programs to address this issue directly. One reviewer of the Working Draft of the *Standards* suggested the establishment of some pilot school mathematics programs based on these *Standards* to demonstrate that *all* students—including women and underserved minorities—can reach a satisfactory level of mathematical achievement and urged that the success of these students be widely publicized.

♦ ♦ ♦ ♦ ♦ ♦ ♦ ♦

Working Conditions. In too many schools, teachers will find it difficult to teach the mathematical topics or create the instructional environments envisioned in these standards because of local constraints, such as directives about which chapters or pages to cover, inadequate time for instruction, and the administration of tests. In many grades too little time is spent on mathematics instruction. Teachers and students should spend an hour a day on mathematics at all grades and take advantage of the many opportunities to connect mathematics to other school subjects.

Teachers also lack the necessary resources, the time to reflect, and the opportunities to share ideas with other teachers. Under such conditions, it is difficult to create a sense of exploration, curiosity, or excitement in the classroom. Although new standards alone cannot alter these conditions, they implicitly argue for everyone to make the work environment for teachers support professional activities.

Research. The *Standards* is based on a set of values, or philosophical positions, about mathematics for students and the way instruction should proceed. These values both are consistent with current research findings and establish a new research agenda. In the redesign of school mathematics, much careful research is needed. Instead of dealing solely with the study of what *is* happening in the teaching and assessment of mathematics instruction, research should deal more with what *ought to be*. For example, the *Standards* offers curricular and pedagogical support for students as they engage in mathematical thinking and problem solving. Although considerable research has dealt with mathematical problem solving, very little of it has examined some of the main components of problem solving, such as conjecturing and problem formulation, described here. Therefore, an examination of these more generative aspects of mathematical thinking is needed.

In summary, our system of schooling needs to be redesigned. We challenge all readers to act, examine these and other constraints of the schooling system, and work toward aligning them with the *Standards*.

Concluding Comments

Today, what happens in America's classrooms is being given lots of attention and scrutiny. The reactions and responses to the recent reports on education offer the mathematics education community a rare opportunity to shape school mathematics during the next decade. Public interest and concern, when combined with changing technology and a growing body of research-based knowledge, are the ingredients necessary for real reform. The NCTM *Standards* is a vehicle that can serve as a basis for improving the teaching and learning of mathematics in America's schools.

The reactions to the Working Draft of the *Standards* have convinced us that many educators are eager to reform school mathematics. Through their professional organization, NCTM, which best reflects their interests and the mathematical learning of their students, knowledgeable teachers and other mathematics educators should assume responsibility for leading the reform effort.

There are, of course, barriers to the implementation of these standards, the most important being the strongly held beliefs, expectations, and attitudes of all people in education about specific aspects of the reform. A teacher who believes that speed in paper-and-pencil calculation is most important will be reluctant to let children use calculators. The administrator who has charted group scores on a standardized test for years will be reluctant to replace it. Parents who expect students to do mathematics homework on paper at a desk rather than by gathering real data to solve a problem will be surprised. The best way to bring about reform is to challenge directly the perceptions held by many about the content of mathematics, what is important for students to learn, the job of teaching, what constitutes the work of students, and the professional roles and responsibilities of teachers and administrators. It is all too easy to agree with the rhetoric of reform but still maintain long-held beliefs or practices inconsistent with intended reform practices. Likewise, it is easy to agree but at the same time claim that it "won't work here." We challenge readers to recognize their beliefs and practices and test them against the standards we have proposed.

Another barrier relates to the political framework within which schools operate. Policy decisions about schooling are made in the context of pressure, consensus, conflict, and compromise by well-meaning elected representatives at the federal, state, provincial, and local levels. These decisions are then made operational by administrative directives. Many of these standards can be fully implemented only by changing directives about the selection of texts, mandated testing, and so forth, in consultation with professionals in mathematics and mathematics education. We challenge policy makers to change the rules.

Still another barrier to reform is cost. Excellence costs money. Most schools, like the communities they serve, are surviving but not thriving. To be successful, any reform requires a considerable commitment of time and resources, and our proposals for school mathematics are no exception. Many resources are scarce, yet they must be found and used judiciously.

These and other barriers to change can be viewed as insurmountable or as challenges to be met and overcome. The *Standards* was produced by working groups confident that if the recommendations are followed, a new school mathematics program can be developed and implemented. The content that should be included in a school mathematics program has been specified. Such materials as texts, courseware, and tests can be produced so that constructive learning will take place in classrooms. However, let it be understood that we hold no illusions of immediate reform. We believe a new program can be developed, but it will be accomplished only by hard work. Our hope and expectations are that a sufficient number of persons are willing to work to accomplish the reform.

Now that you have read this document and deliberated on its vision and recommendations, we reemphasize the following points:

The National Council of Teachers of Mathematics has created a vision of—

- mathematical power for all in a technological society;
- mathematics as something one does—solve problems, communicate, reason;
- a curriculum for all that includes a broad range of content, a variety of contexts, and deliberate connections;
- the learning of mathematics as an active, constructive process;
- instruction based on real problems;
- evaluation as a means of improving instruction, learning, and programs.

♦ ♦ ♦ ♦ ♦ ♦ ♦ ♦

If we keep these points in mind, collectively we have a rare opportunity to provide the kind of leadership that will make real, substantive changes in school mathematics. These changes will ensure that all students possess both a suitable and a sufficient mathematical background to be productive citizens in the next century.

♦ ♦ ♦ ♦ ♦ ♦ ♦ ♦

REFERENCES

Conference Board of the Mathematical Sciences. *The Mathematical Sciences Curriculum K–12: What Is Still Fundamental and What Is Not.* Report to the NSB Commission on Precollege Education in Mathematics, Science, and Technology. Washington, D.C.: CBMS, 1983a.

———. *New Goals for Mathematical Sciences Education.* Report of a conference sponsored by CBMS, Airlie House, Warrenton, Virginia, November 1983. Washington, D.C.: CBMS, 1983b.

Cuban, Larry. "A Fundamental Puzzle of School Reform." *Phi Delta Kappan* 69, no. 5 (1988): 341–44.

Davis, P. J., and R. Hersh. *The Mathematical Experience.* Boston: Houghton Mifflin Co., 1981.

Day, R. P., and R. Scott. "For Problem Solving: Consider Using a Spreadsheet." *Math Times Journal* 1, no. 2 (1987): 20–28.

Dewey, John. *Democracy and Education.* New York: Macmillan Co., 1916.

Freudenthal, Hans. *Mathematics as an Educational Task*, p. 403. Dordrecht, Netherlands: D. Reidel Publishing Co., 1973.

Hirsch, Christian R., ed. *The Secondary School Mathematics Curriculum.* 1985 Yearbook of the National Council of Teacher of Mathematics. Reston, Va.: NCTM, 1985.

Huff, D., and I. Greise. *How to Take a Chance.* New York: W. W. Norton & Co., 1959.

Johnston, W. B., and A. E. Packers. *Workforce 2000: Work and Workers for the Twenty-first Century.* Indianapolis: Hudson Institute, 1987.

Lewis, A. C. "Reading the Writing on the Wall." *Phi Delta Kappan* 69, no. 7 (1988): 468–69.

Ling, John. "SMP 11–16: The Most Recent Work in Curriculum Development by the School Mathematics Project, and Its Relation to Current Issues in Mathematical Education in England." In *Developments in School Mathematics around the World*, edited by Izaak Wirszup and Robert Streit, pp. 233–34. Reston, Va.: National Council of Teachers of Mathematics, 1987.

Meyer, C., and T. Sallee. *Make It Simpler, a Practical Guide to Problem Solving in Mathematics*, p. 241. Menlo Park, Calif.: Addison-Wesley Publishing Co., 1983.

National Commission on Excellence in Education. *A Nation at Risk: The Imperative for Educational Reform.* Washington, D.C.: U.S. Government Printing Office, 1983.

National Council of Teachers of Mathematics. *An Agenda for Action: Recommendations for School Mathematics of the 1980s.* Reston, Va.: NCTM, 1980.

———. "Mathematics for Language Minority Students." *News Bulletin* 23 (May 1987): 9.

National Science Board Commission on Precollege Education in Mathematics, Science, and Technology. *Educating Americans for the Twenty-first Century: A Plan of Action for Improving the Mathematics, Science and Technology Education for All American Elementary and Secondary Students So That Their Achievement Is the Best in the World by 1995.* Washington, D.C.: National Science Foundation, 1983.

Office of Technology Assessment. *Technology and the American Transition.* Washington, D.C.: U.S. Government Printing Office, 1988.

Pólya, George. "On Solving Mathematical Problems in High School." In *Problem Solving in School Mathematics*, 1980 Yearbook of the National Council of Teachers of Mathematics, edited by Stephen Krulik, pp. 1–2. Reston, Va.: NCTM, 1980.

Pollak, Henry. Notes from a talk given at the Mathematical Sciences Education Board. Frameworks Conference, May 1987, at Minneapolis.

Resnick, Lauren B. *Education and Learning to Think*. Washington, D.C.: National Academy Press, 1987.

Romberg, Thomas A. *School Mathematics: Options for the 1990s. Chairman's Report of a Conference*. Washington, D.C.: U.S. Government Printing Office, 1984.

Romberg, Thomas A., and Thomas P. Carpenter. "Research on Teaching and Learning Mathematics: Two Disciplines of Scientific Inquiry." In *Handbook of Research on Teaching: A Project of the American Educational Research Association*, 3d ed., edited by M. C. Wittrock, pp. 850–73. New York: Macmillan Co., 1986.

Schoen, Harold L. "NCTM 9–12 Standards: Reflections from Research." Paper presented at the NCTM/AERA-SIG Research Presession to the NCTM 66th Annual Meeting, April 1988, Chicago.

Steen, Lynn. "A Time of Transition: Mathematics for the Middle Grades." In *A Change in Emphasis*, edited by Richard Lodholz, pp. 1–9. Parkway, Mo.: Parkway School District, 1986.

Swan, Malcolm. *The Language of Functions and Graphs.* Nottingham, England: Shell Centre for Mathematics Education, 1987.

Vergnaud, Gérard. "Multiplicative Structures." In *Number Concepts and Operations in the Middle Grades*, edited by James Hiebert and Merlyn Behr, pp. 141–61. Reston, Va.: National Council of Teachers of Mathematics; Hillsdale, N.J.: Lawrence Erlbaum Associates, 1988.